Selected Titles in This Series

A Gentle Introduction to
GAME THEORY

MATHEMATICAL WORLD • VOLUME 13

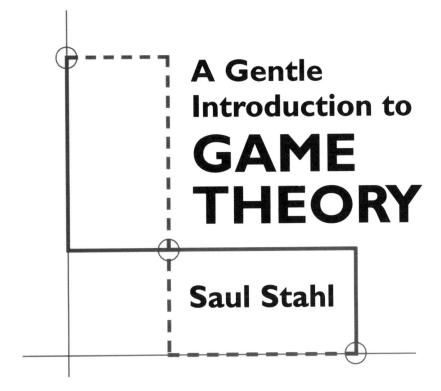

A Gentle Introduction to
GAME THEORY

Saul Stahl

American Mathematical Society

1991 *Mathematics Subject Classification.* Primary 90D05, 90D80.

ABSTRACT. This is an introduction to the theory of 2 person games. Starting from some easy 2×2 zero-sum games, the reader is lead to proofs of the 2×2 cases of von Neumann's Minimax Theorem and the existence of the Nash Equilibrium. Many examples are discussed along the way.

Library of Congress Cataloging-in-Publication Data

Stahl, Saul.
    A gentle introduction to game theory / Saul Stahl.
       p. cm. — (Mathematical world ; v. 13)
    Includes bibliographical references and index.
    ISBN 0-8218-1339-0 (alk. paper)
    1. Game theory.   I. Title.   II. Series.
QA269.S695   1998
519.3—dc21

                                                       98-37248
                                                           CIP

To Dan, Lynne, and Amy

TABLE OF CONTENTS

TABLE OF CONTENTS

PREFACE

Game theory sheds a light on many aspects of the social sciences and is based on an elegant and non-trivial mathematical theory. The bestowal of the 1994 Nobel Prize in economics upon the mathematician John Nash underscores the important role this theory has played in the intellectual life of the twentieth century. There are many textbooks on this topic but they tend to be one sided in their approaches. Some focus on the applications and gloss over the mathematical explanations while others explain the mathematics at a level that makes them inaccessible to most non-mathematicians. This monograph fits in between these two alternatives. Many examples are discussed and completely solved with tools that require no more than high school algebra. These tools turn out to be strong enough to provide proofs of both von Neumann's Minimax Theorem and the existence of the Nash Equilibrium in the 2×2 case. The reader therefore gains both a sense of the range of applications and a better understanding of the theoretical framework of two deep mathematical concepts.

This book is based on lectures I presented in **MATH 105 Introduction to Topics in Mathematics** as well as in **MATH 530 Mathematical Models I** at the University of Kansas. The first of these courses is normally taken by Liberal Arts majors to satisfy their Natural Sciences and Mathematics Distribution Requirements. The presentation of Chapters 1-9 and 11-13 in this class took about 25 lectures and was supplemented with notes on statistics, linear programming and/or symmetry. In the mathematical models class this material was used to supplement a standard linear programming course. It can be covered in about a dozen lectures with proofs included in both the presentation and the homework assignments. Those chapters and exercises that are deemed to be more theoretically demanding are starred. Such proofs as are included in the text appear in the conclusion of the appropriate chapters.

The exposition is *gentle* because it requires only some knowledge of coordinate geometry, and linear programming is *not* used. It is *mathematical* because it is more concerned with the mathematical solution of games than with their applications. Nevertheless, I have included as many convincing applications as I could find.

I am indebted to my colleagues James Fred McClendon for helping me out with some of the technical aspects of the material and Margaret Bayer for rooting out some of the errors in an earlier draft. I owe my own understanding of the material and many examples to the books by John D. Williams and David Gale. David Bitters, Sergei Gelfand, Edward Dunne, and an anonymous reviewer contributed many valuable suggestions. Larisa Martin and Sandra Reed converted the manuscript to TeX and Sandra Donnelly supervised the production. To them all I owe a debt of gratitude.

Saul Stahl

INTRODUCTION

The notion of a zero-sum game is informally introduced and several examples are discussed.

The mathematical theory of games was first developed as a model for situations of conflict. It gained widespread recognition in the early 1940's when it was applied to the theoretical study of economics by the mathematician John von Neumann and the economist Oskar Morgenstern in their book *Theory of Games and Economic Behavior*. Since then its scope has been broadened to include co-operative interactions as well and it has been applied to the theoretical aspects of many of the social sciences. While the jury is still out on the question of whether this theory furnishes any valuable information regarding practical situations, it has stimulated much basic research in disciplines such as economics, political science, and psychology.

Situations of conflict, or any other kind of interactions, will be called *games* and they have, by definition, participants who are called *players*. We shall limit our attention to scenarios where there are only two players and they will be called Ruth and Charlie. The existence of a conflict is usually due both to the desire of each player to improve his circumstances, frequently by means of some acquisition, and the unfortunate limited nature of all resources. For all but the last three chapters of this book it will be assumed that each player is striving to *gain* as much as possible, and that each player's gain is his opponent's loss. Finally, each player is assumed to have several *options* or *strategies* that he can exercise (one at a time) as his attempt to claim a portion of the resources. Because of the introductory nature of this text, most of the subsequent discussion is restricted to situations wherein the players make their moves simultaneously and independently of each other. It will be argued in Chapter 10 that this does not truly limit the scope of the theory and that the mathematical theory of games does have something to say about games, such as poker, in which the players move alternately and do possess a fair amount of information about their opponent's actions.

The foregoing discussion is admittedly vague and this chapter's remainder is devoted to the informal exposition of several examples of increasing complexity. The requisite formal definitions are postponed to the next chapter.

PENNY-MATCHING. *Ruth and Charlie each hold a penny and they display them simultaneously. If the pennies match in the sense that both show heads or both show tails, then Ruth collects both coins. Otherwise, Charlie gets them.*

It is clear that this game is fair in the sense that neither player has an advantage over the other. Moreover, if the game is only played once, then neither player possesses a shrewd system, or strategy, that will improve his position, nor is there a foolish decision that will worsen it.

The situation changes if the game is played many times. While the repeated game remains symmetrical, and neither player has a strategy that will guarantee his coming out ahead in the long run, it is possible to play this game foolishly. Such would be the case were Ruth to consistently display the *head* on her penny. In that case Charlie would be sure to catch on and display the *tail* each time, thus coming out ahead. It is a matter of common sense that neither player should make any predictable decisions, nor should a player favor either of his options. In other words, when this game is repeated many times each player should play each of his options with equal frequency 1/2 and make his decisions unpredictable. One way a player can accomplish this is by flipping his coin rather than consciously deciding which side to show. With an eye to the analysis of more complex games, this game is summarized as

| | **Charlie** | |
	Head	Tail
Ruth Head	1	-1
Tail	-1	1

The entry 1 denotes a gain of one penny for Ruth, and the entry -1 denotes a loss of one penny for Ruth. Since Ruth's gain is Charlie's loss, and vice versa, this array completely describes the various outcomes of a single play of the game.

All the subsequent examples and most of the discussion will be phrased in terms of the same two players Ruth and Charlie. Payoffs will be described from Ruth's point of view. Thus, a payoff of a penny will always mean a penny gained by Ruth and lost by Charlie.

ROCK-SCISSORS-PAPER. *Ruth and Charlie face each other and simultaneously display their hands in one of the following three shapes: a fist denoting a rock, the forefinger and middle finger extended and spread to as to suggest scissors, or a downward facing palm denoting a sheet of paper. The rock wins over the scissors since it can shatter them, the scissors win over the paper since they*

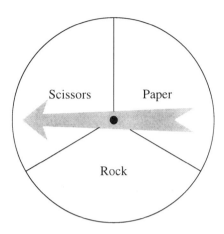

FIGURE 1.1. A randomizing spinner.

can cut it, and the paper wins over the rock since it can be wrapped around the latter. The winner collects a penny from the opponent and no money changes hands in the case of a tie.

This game has features that are very similar to those of Penny-matching. There are neither shrewd nor foolish decisions for a single play. If the game is repeated many times then players who favor one of the options place themselves at a disadvantage. The best strategy for each player is to play each of the options with the same frequency of 1/3 in a manner that will yield the opponent as little information as possible about any particular decision. For example, he could base each of his decisions on the result of a spin of the spinner of Figure 1.

The outcomes of the Rock-scissors-paper game are tabulated as

| | | **Charlie** | | |
		Rock	Scissors	Paper
	Rock	0	1	−1
Ruth	Scissors	−1	0	1
	Paper	1	−1	0

As before, a positive entry denotes a gain for Ruth whereas a negative entry is a gain for Charlie.

The biologists B. Sinervo and C. M. Lively have recently reported on a lizard species whose males are divided into three classes according to their mating behavior. The interrelationship of these three alternative behaviors very much resembles the Rock-Scissors-Paper game and merits a digression here. Each male of the side-blotched lizards (*Uta stansburiana*) exhibits one of three (genetically transmitted) mating behaviors:

a) *highly aggressive* with a large territory that includes several females;

b) *aggressive* with a smaller territory that holds one female;

c) nonagressive *sneaker* with no territory who copulates with the others' females.

In a confrontation, the highly aggressive male has the advantage over the monogamous one who in turn has the advantage over the sneaker. However, because the highly aggressive males must split their time between their various consorts, they are vulnerable to the sneakers. The observed consequence of this is that the male populations cycle from a high frequency of aggressives to a high frequency of highly aggressives, then on to a high frequency of sneakers and back to a high frequency of aggressives.

The next game is played in Italy using three fingers. For pedagogical reasons it is its two finger simplification that will be examined here.

Two-Finger Morra. *At each play Ruth and Charlie simultaneously extend either one or two fingers and call out a number. The player whose call equals the total number of extended fingers wins that many pennies from the opponent. In the event that neither players' call matches the total, no money changes hands.*

It would be clearly foolish for a player to call a number that cannot possibly match the total number of displayed fingers. Thus, a player who extends only one finger would call either 2 or 3, whereas a player who extends two fingers would call only 3 or 4. Consequently, each player has in reality only four options and the game's possible outcomes are summarized as

Charlie

		(1,2)	(1,3)	(2,3)	(2,4)
	(1,2)	0	2	−3	0
	(1,3)	−2	0	0	3
Ruth	(2,3)	3	0	0	−4
	(2,4)	0	−3	4	0

where the option (i, j) denotes the extension of i fingers and a call of j.

This game, like its predecessors, is symmetrical. Neither player has a built-in advantage. It is tempting, therefore, to conclude that when this game is played many times, each player should again randomize his decisions and play them each with a frequency of 1/4 . This, however, turns out to be a poor strategy. If Ruth did randomize this way then Charlie could ensure his coming out ahead in the long run by consistently employing option (1, 3). Note that in that case the expected outcome per play can be computed by the rules of probability (see the discussion at the end of this chapter) as

$$\frac{1}{4} \times 2 + \frac{1}{4} \times 0 + \frac{1}{4} \times 0 + \frac{1}{4} \times (-3) = -\frac{1}{4}.$$

In other words, if Ruth plays each option with a frequency of 1/4 and Charlie consistently employs option (1, 3), Charlie can expect to win, on the average, 1/4 of a penny per play. Thus, this game differs drastically from the two previous ones. Whereas the symmetry of the game makes it clear that each player should be able to pursue a long term strategy that entitles them to expect to come out more or less even in the long run, it is not at all clear this time what this strategy is. We shall return to this game and issue in Chapter 11.

The above considerations point out a difficulty that will have to be dealt with when the general theory of games is proposed. It was noted that if Ruth randomizes her behavior by using each of her four options with a frequency of 1/4, then Charlie can guarantee a long run advantage by consistently employing option (1, 3). However, should Charlie choose to do so, Ruth is bound to notice his bias, and she will in all likelihood respond by consistently opting for (1, 2), yielding her a win of 2 on each play. Charlie, then, will respond by playing (2, 3) consistently for a win of 3 on each play. Ruth, then, will switch to (2, 4) for a repeated gain of 4. Charlie, then, will switch back to (1, 3), thus beginning the whole cycle again. A reasonable theory of games should provide a stable strategy that avoids such "logical loops", and we will see that such is indeed the case.

BOMBING SORTIES. *Ruth and Charlie are generals of opposing armies. Every day Ruth sends out a bombing sortie that consists of a heavily armed bomber plane and a lighter support plane. The sortie's mission is to drop a single bomb on Charlie's forces. However, a fighter plane of Charlie's is waiting for them in ambush and it will dive down and attack one of the planes in the sortie once. The bomber has an 80% chance of surviving such an attack, and if it survives it is sure to drop the bomb right on the target. General Ruth also has the option of placing the bomb on the support plane. In that case, due to this plane's lighter armament and lack of proper equipment, the bomb will reach its target with a probability of only 50% or 90%, depending on whether or not it is attacked by Charlie's fighter.*

This information is summarized in the table below where the entries denote the probability of the bomb's delivery.

		Attack	
		Bomber	Support
Bomb placement	Bomber	80%	100%
	Support	90%	50%

Ruth knows that if the bomb is placed consistently on the bomber she can reasonably expect at least 80% of the missions to succeed. In all likelihood, Charlie's observers at the bombing site would notice this bias and he would direct his fighter plane pilot to always attack the bomber, thus holding Ruth's expectation down to 80% and no more. However, Ruth, who is an experienced poker player, decides to bluff by placing the bomb on the support plane occasionally. Let us

Table 1.1 Bombing sorties.

Compound event	Probability of event	Likelihood of success of sortie
The bomb is on the bomber and *the fighter attacks the bomber*	$.75 \times .50 = .375$	80%
The bomb is on the bomber and *the fighter attacks the support plane*	$.75 \times .50 = .375$	100%
The bomb is on the support plane and *the fighter attacks the bomber*	$.25 \times .50 = .125$	90%
The bomb is on the support plane and *the fighter attacks the support plane*	$.25 \times .50 = .125$	50%

say for the sake of argument that she does so $1/4 = .25$ of the time. Charlie now faces a dilemma. His observers have advised him of Ruth's new strategy and he suspects that it would be advantageous for him to attack the support plane some of the time, but how often should he do so?

Suppose Charlie decides to counter Ruth's bluffing by attacking the support plane half the time. In this case the situation is summarized as

			Attack frequencies	
			.50	.50
			Bomber	Support
Bomb	.75	Bomber	80%	100%
placement frequencies	.25	Support	90%	50%

Since Ruth and Charlie make their daily decisions independently of each other, it follows that on any single sortie, the probabilities of each of the four possible outcomes occurring are as displayed in Table 1 above. Hence, under these circumstances, wherein the bomb is placed on the support plane $1/4$ of the time and this plane is attacked by the fighter $1/2$ of the time, the percentage of successful missions, as computed on the basis of Table 1, is

$$.375 \times 80\% + .375 \times 100\% + .125 \times 90\% + .125 \times 50\%$$
$$= 30\% + 37.5\% + 11.25\% + 6.25\% = 85\%.$$

Thus, this response of Charlie's to Ruth's bluffing has created a situation wherein the bomb can be expected to get through 85% of the time. Since this figure amounts to only 80% when the bomb is placed consistently in the bomber, Ruth's bluffing seems to have paid off.

Table 1.2 Bombing sorties.

Compound event	Probability of event	Likelihood of success of sortie
The bomb is on the bomber and *the fighter attacks the bomber*	$.75 \times .80 = .60$	80%
The bomb is on the bomber and *the fighter attacks the support plane*	$.75 \times .20 = .15$	100%
The bomb is on the support plane and *the fighter attacks the bomber*	$.25 \times .80 = .20$	90%
The bomb is on the support plane and *the fighter attacks the support plane*	$.25 \times .20 = .05$	50%

However, Charlie can change his response pattern. He could, say, decide to diminish the frequency of attacks on the support plane to only 1/5 of the time, leading to the situation

			Attack frequencies	
			.80	.20
			Bomber	Support
Bomb placement frequencies	.75	Bomber	80%	100%
	.25	Support	90%	50%

In this case the percentage of successful sorties, computed on the basis of Table 2, is

$$.60 \times 80\% + .15 \times 100\% + .20 \times 90\% + .05 \times 50\% = 48\% + 15\% + 18\% + 2.5\% = 83.5\%.$$

Thus, by diminishing the frequency of attacks on the support plane Charlie would create a situation wherein only 83.5% of the sorties would be successful. From Charlie's point of view this is an improvement on the 85% computed above. Could greater improvements be obtained by a further diminution of the frequency of attacks on the support plane? Suppose these attacks are completely eliminated. Then the state of affairs is

			Attack frequencies	
			1	0
			Bomber	Support
Bomb placement frequencies	.75	Bomber	80%	100%
	.25	Support	90%	50%

Table 1.3 Bombing sorties.

Compound event	Probability of event	Likelihood of success of sortie
The bomb is on the bomber and *the fighter attacks the bomber*	$.75 \times 1 = .75$	80%
The bomb is on the bomber and *the fighter attacks the support plane*	$.75 \times 0 = 0$	100%
The bomb is on the support plane and *the fighter attacks the bomber*	$.25 \times 1 = .25$	90%
The bomb is on the support plane and *the fighter attacks the support plane*	$.25 \times 0 = 0$	50%

and the associated Table 3 tells us that the percentage of successful sorties is

$$.75 \times 80\% + 0 \times 100\% + .25 \times 90\% + 0 \times 50\%$$
$$= 60\% + 0\% + 22.5\% + 0\% = 82.5\%.$$

This would seem to indicate that when Ruth bluffs 1/4 of the time Charlie should nevertheless ignore the support plane and direct his attacks exclusively at the bomber. Moreover, these calculations indicate that this strategy enables Ruth to improve on her bomber's 80% delivery rate by another 2.5%. These various analyses raise several questions that will guide the development of the subsequent sections. Can Ruth improve on the above 82.5% with a different strategy? What is the best improvement Ruth can obtain? What is Charlie's best response to any specific strategy of Ruth's? Does Charlie have an overall best strategy that is independent of Ruth's decisions?

An Addendum on Probabilistic Matters

Unless otherwise stated, it will always be assumed here that each time a game is played Ruth and Charlie make their respective decisions independently of each other. This comes under the general heading of *Independent Random Events*; i.e., random events whose outcomes have no bearing on each other. For example, if a nickel and a dime are tossed the outcomes will in general be independent, unless the two coins are glued to each other. Similarly, suppose Ruth draws a card at random from a standard deck, replaces it, shuffles the deck, and then Charlie draws a card from that deck. The two draws are then independent. On the other hand, if Ruth does not replace her card in the deck, then Charlie's draw is very much affected by Ruth's draw, since he cannot possibly draw the same card as she did. In this case the two random draws are not independent. The probabilities of independent events are related by the formula

If the two events E and F are independent then

probability of (E and F) = (probability of E) × (probability of F).

Thus, the probability of the aforementioned coins both coming up *heads* is $.5 \times .5 = .25$. Similarly, If a nickel and a standard six sided die are tossed simultaneously, then the probability of the coin coming up *tails* and the die showing a 3 is $\frac{1}{2} \cdot \frac{1}{6} = \frac{1}{12}$.

Some random events have numerical values attached to their outcomes. Thus, the faces of the standard die have dots marked on them, a person selected at random has a height, and a lottery ticket has a monetary value (that is unknown at the time it is purchased). Loosely speaking, the *expected value* is the average of those numerical values when they are weighted by the corresponding probabilities. More formally:

If the random variable X assumes the numerical values $x_1, x_2, \ldots, x_n$ with probabilities $p_1, p_2, \ldots, p_n$ respectively, then the expected value (weighted average) of X is

$$p_1 x_1 + p_2 x_2 + \cdots + p_n x_n.$$

For example, suppose that the records of an insurance company indicate that during a year they will pay out for accidents according to the following pattern:

$$\$100,000 \quad \text{with probability } .0002$$
$$\$50,000 \quad \text{with probability } .0015$$
$$\$25,000 \quad \text{with probability } .003$$
$$\$5,000 \quad \text{with probability } .01$$
$$\$1,000 \quad \text{with probability } .03$$
$$\$0 \quad \text{with probability } .9553$$

Then the expected payment per car during the next year is

$$\$100,000 \cdot .0002 + \$50,000 \cdot .0015 + \$25,000 \cdot .003 + \$5,000 \cdot .01 + \$1,000 \cdot .03$$
$$+ \$0 \cdot .9553 = \$250.$$

Similarly, if 5000 lottery tickets are sold of which one will win \$10,000, two will win \$1,000, five will win \$100 and the rest will all receive a consolation prize worth one dime, then the expected value of each ticket is

$$\$10,000 \cdot \frac{1}{5000} + \$1,000 \cdot \frac{2}{5000} + \$100 \cdot \frac{5}{5000} + \$.10 \cdot \frac{4992}{5000} \approx \$2.60.$$

Chapter Summary

Four examples of repeated games were discussed. For the first two, Penny-matching and Scissors-paper-rock, the players have obvious reasonable goals: to come out even. The means for attaining these goals are also clear: they should randomize their decisions and employ their options with equal frequencies. In the case of Morra, the players are again entitled to expect to come out even in the long run, but the means of attaining this goal are not apparent. The player who randomizes his Morra options and plays them with equal frequencies places himself at a disadvantage. Finally, in the Bombing-sorties game neither the exact goal nor the means of attaining it are evident.

Chapter Terms

Bombing sorties	5	Conflict	1
Gain	1	Game	1
Interaction	1	Loss	1
Morra	4	Option	1
Payoff	2	Penny-matching	2
Rock-scissors-paper	2	Strategy	1

EXERCISES 1

1. Ruth is playing Morra and has decided to play the options with the following frequencies:

$$(1, 2) - 40\%$$
$$(1, 3) - 30\%$$
$$(2, 3) - 20\%$$
$$(2, 4) - 10\%.$$

   a) Design a spinner that will facilitate this pattern;
   b) What will be the expected outcome if Charlie consistently plays (1,2)?
   c) What will be the expected outcome if Charlie consistently plays (1,3)?
   d) What will be the expected outcome if Charlie consistently plays (2,3)?
   e) What will be the expected outcome if Charlie consistently plays (2,4)?
   f) Which of the above consistent moves is best for Charlie?

2. Ruth is playing Morra and has decided to play the options with the following frequencies:

$$(1, 2) - 0\%$$
$$(1, 3) - 50\%$$
$$(2, 3) - 50\%$$
$$(2, 4) - 0\%.$$

   a) Design a spinner that will facilitate this pattern;
   b) What will be the expected outcome if Charlie consistently plays (1,2)?
   c) What will be the expected outcome if Charlie consistently plays (1,3)?
   d) What will be the expected outcome if Charlie consistently plays (2,3)?
   e) What will be the expected outcome if Charlie consistently plays (2,4)?
   f) Which of the above consistent moves is best for Charlie?

3. Ruth is playing Morra and has decided to play the options with the following frequencies:

$$(1, 2) - 10\%$$
$$(1, 3) - 20\%$$
$$(2, 3) - 30\%$$
$$(2, 4) - 40\%.$$

    a) Design a spinner that will facilitate this pattern;
    b) What will be the expected outcome if Charlie consistently plays (1,2)?
    c) What will be the expected outcome if Charlie consistently plays (1,3)?
    d) What will be the expected outcome if Charlie consistently plays (2,3)?
    e) What will be the expected outcome if Charlie consistently plays (2,4)?
    f) Which of the above consistent moves is best for Charlie?

4. General Ruth has decided that she will put the bomb on the support plane 10% of the time.
    a) Design a spinner that will facilitate this pattern.
    b) What will be the expected mission success rate if General Charlie persists in attacking the bomber exclusively?
    c) What will be the expected mission success rate if General Charlie persists in attacking the support plane exclusively?
    d) What will be the expected mission success rate if General Charlie attacks each plane 50% of the time?
    e) What will be the expected mission success rate if General Charlie attacks the support plane 40% and the bomber 60% of the time?

5. General Ruth has decided that she will put the bomb on the support plane 30% of the time.
    a) Design a spinner that will facilitate this pattern.
    b) What will be the expected mission success rate if General Charlie persists in attacking the bomber exclusively?
    c) What will be the expected mission success rate if General Charlie persists in attacking the support plane exclusively?
    d) What will be the expected mission success rate if General Charlie attacks each plane 50% of the time?
    e) What will be the expected mission success rate if General Charlie attacks the support plane 40% and the bomber 60% of the time?

6. General Ruth has decided that she will put the bomb on the support plane 40% of the time.
    a) Design a spinner that will facilitate this pattern.
    b) What will be the expected mission success rate if General Charlie persists in attacking the bomber exclusively?
    c) What will be the expected mission success rate if General Charlie persists in attacking the support plane exclusively?
    d) What will be the expected mission success rate if General Charlie attacks each plane 50% of the time?
    e) What will be the expected mission success rate if General Charlie attacks the support plane 40% and the bomber 60% of the time?

THE FORMAL DEFINITIONS

**The fundamental notions of a zero-sum game,
mixed and pure strategies, and expected payoff are defined.**

It is time to make some formal definitions. For purely pedagogical reasons we begin with games in which each of the players has only two options, leaving the more general case for the end of the section.

Taking our cue from the Penny-matching and Bombing-sorties games of the previous section, we define a 2×2 *zero-sum game* as a square array of $2 \times 2 = 4$ numbers. Thus, the mathematical representation of the Penny-matching game is the array

1	−1
−1	1

and the *mathematical* representation of the Bombing-sorties game is the array

80%	100%
90%	50%

Just as it is convenient to represent the addition of 8 oranges to 6 oranges by the abstract equation $8 + 6 = 14$, we shall ignore the actual details of the games in most of the subsequent discussion and simply deal with arrays of unitless numbers. Thus, the most general 2×2 zero-sum game has the form

a	b
c	d

where a, b, c, d are arbitrary numbers. This abstraction has the advantages of succinctness and clarity. We shall, however, make a point of discussing some concrete games every now and then, and many more such games will be found in the exercises.

Each 2×2 zero-sum game has two players, whom we shall continue to call Ruth and Charlie. The mathematical analog of deciding on one of the options is the selection of either a row or a column of this array. Specifically, Ruth decides on an option by selecting a row of the array, whereas Charlie makes his decision by specifying a column. Thus, in Bombing-sorties, Ruth's placement of the bomb in the bomber, and Charlie's attacking the support plane are tantamount to Ruth's selecting the first row and Charlie's selecting the second column of the array

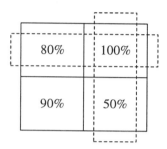

Each individual *play* of the game consists of such a pair of selections, made simultaneously and independently. The selected row and column constitute the *outcome* of the play and its *payoff* is the entry of the array that is contained in both of the selections. Thus, the payoff of the play illustrated above is 100%. On the other hand, had Ruth selected the second row and had Charlie stayed with the second column the payoff would have been 50%. This payoff of course represents Ruth's winnings (and Charlie's loss) from that play and she will in general wish to maximize its value, whereas Charlie will be guided by the desire to minimize this payoff.

Informally speaking, a players' strategy is a decision on the frequency with which each available option will be chosen. More formally, a *strategy* is a pair of numbers $[a, b]$,

$$0 \leq a \leq 1, \quad 0 \leq b \leq 1, \quad a + b = 1,$$

where a denotes the frequency with which the first row (or column) is chosen, and b denotes the frequency with which the second row (or column) is chosen. Thus, Ruth's decision to place the bomb on the support plane $1/4$ of the time is denoted by the pair $[.75, .25]$ and Charlie's strategy of attacking the two planes with equal frequencies is denoted by $[.5, .5]$. When discussing general strategies it is convenient to denote Ruth's strategy by $[1 - p, p]$ and Charlie's strategy by $[1 - q, q]$. The rationale for placing the p and the q in the second component rather than the first one will be explained in Chapter 4.

This formalization of the intuitive concept of strategy as a list of probabilities was first offered by the mathematician E. Borel in series of papers that were written in the 1920's. Borel was also the first one to view these games as rectangular arrays. The decision to limit the notion of strategy in this manner could not have been an easy one and Borel's papers exhibit some ambivalence on this issue. It is very tempting to believe, as did Borel, that one can gain some advantage over the opponent by varying one's probabilities at each play, but it is not at all clear how to formulate such a variation. At the conclusion of one of his papers he wrote:

> The function $f(x, y)$ [the strategy] must then vary at each instant, and vary without following any law at all. One may well doubt if it is possible to indicate an effective and sure means of carrying out such counsel. It seems that, to follow it to the letter, a complete incoherence of mind would be needed, combined, of course, with the intelligence necessary to eliminate those methods we have called bad.

It is our contention, and assumption, that, in practice, every player has a strategy in the sense of a fixed pair of probabilities. Truly random behavior is impossible to attain within the context of a game. For how can any person repeatedly play a game such as Rock-scissors-paper in a truly random fashion, that is to say, without some pattern of frequencies emerging from his choices? If he makes each individual decision in his head, his past experience and personal preferences are sure to dictate a pattern. If, on the other hand, he uses some device such as a die or a spinner to implement the randomization, the device itself will shape the randomness into a frequency distribution. Each face of the die will come up approximately 1/6 of the time and the spinner's arrow will stop within each sector with a frequency that is proportional to that sector's central angle.

Another argument against the feasibility of a strategyless manner of playing is that each actual sequence of choices made by a player over his lifetime can be construed as a strategy. If a player of Penny-matching has been observed to have displayed Heads 83 times and Tails 39 times, it could be reasonably said that, since $83 + 39 = 122$, it follows that this player has the strategy

$$\left[\frac{83}{122}, \frac{39}{122} \right] \approx [.68, .32]$$

In reality, of course, such statistics are rarely available. However, the point we are making here is that it is permissible to assume that *every* player is indeed employing some strategy that may or may not be known to the opponent.

More serious than the inability of people to behave in a truly random fashion is the theoretical impossibility of treating such a concept within a mathematical framework. Even in the mathematical theory of statistics, where random variables are as a commonplace as ants at a picnic, every such variable is assumed to have a probability distribution. Von Neumann and Morgenstern, who also identified strategies with probabilities as was done above, explicitly admitted that they were doing so simply because they had no theory that could handle anything else (see discussion at the end of Chapter 4).

Thus, it will henceforth be assumed that

Every player employs a strategy.

It was seen in Bombing-sorties that the specification of each player's strategy resulted in a situation wherein an expected payoff could be computed. This *expected payoff* can be defined and computed for arbitrary 2×2 games in a similar manner. Thus, given the strategies $[1 - p, p]$ and $[1 - q, q]$ for the general 2×2 zero-sum game

$$
\begin{array}{c|c|c|}
 & 1-q & q \\
\hline
1-p & a & b \\
\hline
p & c & d \\
\hline
\end{array}
\tag{1}
$$

the likelihood of Ruth getting the payoff a is the probability of her choosing the first row and Charlie choosing the first column. Since these choices are made independently, we conclude that

the probability of Ruth getting payoff a is $(1 - p) \times (1 - q)$.

Similarly,

the probability of Ruth getting payoff b is $(1 - p) \times q$

the probability of Ruth getting payoff c is $p \times (1 - q)$

the probability of Ruth getting payoff d is $p \times q$.

As these four events are mutually exclusive and they exhaust all the possibilities it follows that Ruth's expected payoff is

$$(1 - p) \times (1 - q) \times a + (1 - p) \times q \times b + p \times (1 - q) \times c + p \times q \times d.$$

Diagrams such as that of (1) above, wherein the players' strategies are appended to the game's array, are called *auxiliary diagrams*. They turn the computation of the expected payoff into a routine task and so will be repeatedly used in the sequel as a visual aid.

EXAMPLE 1. For the Bombing-sorties game Ruth's strategy $[.3, .7]$ and Charlie's strategy $[.6, .4]$ yield the auxiliary diagram

$$
\begin{array}{c|c|c|}
 & .6 & .4 \\
\hline
.3 & 80\% & 100\% \\
\hline
.7 & 90\% & 50\% \\
\hline
\end{array}
$$

The corresponding payoff is

$$.3 \times .6 \times 80\% + .3 \times .4 \times 100\% + .7 \times .6 \times 90\% + .7 \times .4 \times 50\%$$
$$= 14.4\% + 12\% + 37.8\% + 14\% = 78.2\%.$$

In other words, when the players employ the above specified strategies, Ruth can expect 78.2% of the missions to be successful.

EXAMPLE 2. Compute the expected payoff of the strategies $[.2, .8]$ of Ruth and $[.3, .7]$ of Charlie for the abstract 2×2 zero-sum game

5	0
−1	2

The auxiliary diagram is

	.3	.7
.2	5	0
.8	−1	2

and so the expected payoff is

$$.2 \times .3 \times 5 + .2 \times .7 \times 0 + .8 \times .3 \times (-1) + .8 \times .7 \times 2$$
$$= .3 + 0 - .24 + 1.12 = 1.18.$$

In other words, when the players use the specified strategies within the context of the given (repeated) game, Ruth can expect to win, on the average, 1.18 per play. Of course, this being a zero-sum game, Charlie should expect to lose the same amount per play.

We now turn to the formal definition of games in which one or both of the players have more than two options. If m and n are any two positive integers, then an $m \times n$ *zero-sum game* is a rectangular array of mn numbers having m rows and n columns. Thus,

0	1	−1
−1	0	1
1	−1	0

is a 3×3 zero-sum game (Rock-paper-scissors), and

0	2	−3	0
−2	0	0	3
3	0	0	−4
0	−3	4	0

is a 4×4 zero-sum game (Two-finger Morra). The array below is an example of an abstract 3×4 zero-sum game (an interesting concrete nonsquare game will be discussed in Chapter 7 in detail).

2	−1	−5	3
0	−2	3	−3
1	0	1	−2

$$\text{(2)}$$

It is again assumed that in each play Ruth selects a row and Charlie selects a column of the array. The selected row and column constitute the *outcome* of that play and the entry of the array which is the intersection of Ruth's chosen row with Charlie's chosen column is the corresponding *payoff*. Thus, in the above abstract game, if Ruth selects the second row and Charlie selects the fourth column, the corresponding payoff is −3, a loss for Ruth.

Given an $m \times n$ zero-sum game, a *strategy* for Ruth is a list of numbers $[p_1, p_2, \ldots, p_m]$, such that

$$0 \le p_i \le 1 \qquad \text{for each } i = 1, 2, \ldots, m, \text{ and}$$
$$p_1 + p_2 + \ldots + p_m = 1,$$

where p_i denotes the frequency with which Ruth chooses the i-th row. Similarly, a strategy for Charlie is an ordered list of numbers $[q_1, q_2, \ldots, q_n]$, such that

$$0 \le q_j \le 1 \qquad \text{for each } j = 1, 2, \ldots, n, \text{ and}$$
$$q_1 + q_2 + \ldots + q_n = 1,$$

where q_j denotes the frequency with which Charlie chooses the j-th column. Thus, in the symmetrical Rock-scissors-paper game, the strategy $[.6, .3, .1]$ denotes the decision, by either player, to display *rock* 60% of the time, *scissors* 30% of the time, and *paper* only 10% of the time. In the above abstract 3×4 zero-sum game (or in any 3×4 zero-sum game for that matter) the strategy $[.2, .3, .5]$ denotes Ruth's choosing the first, second, or third rows 20%, 30%, 50% of the time respectively. The strategy $[.4, 0, .1, .5]$ denotes Charlie's choosing the first, second, third, or fourth columns 40%, 0%, 10%, and 50% of the time. A *pure strategy* is one which calls for the exclusive use of a particular row or column. In Bombing-sorties each player has two pure strategies: $[1, 0]$ which calls for the use of the first row or column only, and $[0, 1]$ which calls for the exclusive use of the second row or column. In the abstract game of (2), Ruth has three pure strategies: $[1, 0, 0]$, $[0, 1, 0]$, and $[0, 0, 1]$. In the same game Charlie has four pure strategies: $[1, 0, 0, 0]$, $[0, 1, 0, 0]$, $[0, 0, 1, 0]$, and $[0, 0, 0, 1]$. Strategies which are not known to be pure are called *mixed*. Thus, the strategies $[.3, .7]$ and $[1 - p, p]$ are mixed strategies, even though the later may turn out to be pure when p is either 0 or 1.

Every such choice of specific strategies on the part of both players narrows the situation down to a point where an expected payoff can be computed. Suppose Ruth employs the strategy $[p_1, p_2, \ldots, p_m]$ and Charlie employs the strategy $[q_1, q_2, \ldots, q_n]$ in some abstract $m \times n$ zero-sum game. If $a_{i,j}$ denotes the payoff in the i-th row and the j-th column of this game (see Table 2.1), then the likelihood of this payoff actually taking place is the probability of Ruth choosing

TABLE 2.1. Computing the expected payoff

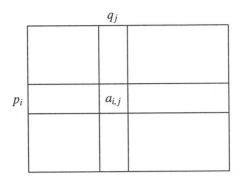

the i-th row and Charlie choosing the j-th column, which, of course, equals $p_i \times q_j$. Thus, this specific outcome's contribution to the expected payoff is $p_i \times q_j \times a_{i,j}$. Consequently the total *expected payoff*, being the sum of all these contributions, equals

Sum of all $p_i \times q_j \times a_{i,j}$ $\quad i = 1, 2, \ldots, m, \; j = 1, 2, \ldots, n.$

EXAMPLE 3. Compute the expected payoff when Ruth employs the strategy $[.2, .3, .5]$ and Charlie employs the strategy $[.1, .7, .2]$ in the Rock-scissors-paper game.

The auxiliary diagram is

	.1	.7	.2
.2	0	1	−1
.3	−1	0	1
.5	1	−1	0

and the corresponding sum is

$$.2 \times .1 \times 0 + .2 \times .7 \times 1 + .2 \times .2 \times (-1)$$
$$+ .3 \times .1 \times (-1) + .3 \times .7 \times 0 + .3 \times .2 \times 1$$
$$+ .5 \times .1 \times 1 + .5 \times .7 \times (-1) + .5 \times .2 \times 0$$
$$= 0 + .14 - .04 - .03 + 0 + .06 + .05 - .35 + 0 = -.17.$$

In other words, under these circumstances, Ruth should expect to lose .17 pennies per play.

EXAMPLE 4. Compute the expected payoff when the strategies $[.3, 0, .7]$ and $[.1, .2, .3, .4]$ are used by Ruth and Charlie respectively in the abstract game (2) above.

Using the auxiliary diagram

	.1	.2	.3	.4
.3	2	−1	−5	3
0	0	−2	3	−3
.7	1	0	1	−2

we get an expected payoff of

$$.3 \times .1 \times 2 + .3 \times .2 \times (-1) + .3 \times .3 \times (-5) + .3 \times .4 \times 3$$
$$+ 0 \times .1 \times 0 + 0 \times .2 \times (-2) + 0 \times .3 \times 3 + 0 \times .4 \times (-3)$$
$$+ .7 \times .1 \times 1 + .7 \times .2 \times 0 + .7 \times .3 \times 1 + .7 \times .4 \times (-2)$$
$$= .06 - .06 - .45 + .36 + 0 + 0 + 0 + 0 + .07 + 0 + .21 - .56 = -.37.$$

Chapter Summary

The notion of an abstract zero-sum two person game was extracted from the concrete examples of the previous section. The associated concepts of strategies and expected payoffs were formally defined.

Chapter Terms

EXERCISES 2

In each of Exercises 1–11 compute the expected payoff where **R** denotes a strategy for Ruth and **C** denotes a strategy for Charlie for the given game **G**.

1. **R** = [.2, .8], **C** = [.7, .3], **G** =

2	3
4	1

2. **R** = [.6, .4], **C** = [0, 1], **G** =

2	3
4	1

3. **R** = [.2, .8], **C** = [.7, .3], **G** =

−1	3
4	−2

4. $\mathbf{R} = [.6, .4]$, $\mathbf{C} = [0, 1]$, $\mathbf{G} =$

−1	3
4	−2

5. $\mathbf{R} = [.6, .3, .1]$, $\mathbf{C} = [.1, .4, .3, .2]$, $\mathbf{G} =$

1	0	−2	3
−3	4	2	−4
0	−1	0	1

6. $\mathbf{R} = [.6, 0, .4]$, $\mathbf{C} = [0, .5, .3, .2]$, $\mathbf{G} =$

1	0	−2	3
−3	4	2	−4
0	−1	0	1

7. $\mathbf{R} = [0, 1, 0]$, $\mathbf{C} = [.5, 0, 0, .5]$, $\mathbf{G} =$

1	0	−2	3
−3	4	2	−4
0	−1	0	1

8. $\mathbf{R} = [.2, 0, .4, 0, .4]$, $\mathbf{C} = [.1, .1, .8]$, $\mathbf{G} =$

1	−3	2
0	4	−4
−2	0	2
3	−3	−1
−3	5	1

9. $\mathbf{R} = [0, 0, 0, .4, .6]$, $\mathbf{C} = [.1, 0, .9]$, $\mathbf{G} =$

1	−3	2
0	4	−4
−2	0	2
3	−3	−1
−3	5	1

1	−3	2
0	4	−4
−2	0	2
3	−3	−1
−3	5	1

10. $\mathbf{R} = [0, 0, 0, 1, 0]$, $\mathbf{C} = [0, 1, 0]$, $\mathbf{G} =$

1	−3	2
0	4	−4
−2	0	2
3	−3	−1
−3	5	1

11. $\mathbf{R} = [0, 1, 0, 0, 0]$, $\mathbf{C} = [0, 0, 1]$, $\mathbf{G} =$

The following games are reprinted with the kind permission of the RAND Corporation. Describe each as a table.

12. **The River Tale** (J. D. Williams) Steve is approached by a stranger who suggests that they match coins. Steve says that it's too hot for violent exercise. The stranger says, "Well then, let's just lie here and speak the words *heads* or *tails*—and to make it interesting I'll give you $30 when I call *tails* and you call *heads*, and $10 when it's the other way round. And—just to make it fair—you give me $20 when we match."

13. **The Birthday** (J. D. Williams) Frank is hurrying home late, after a particularly grueling day, when it pops into his mind that today is Kitty's birthday! Or is it? Everything is closed except the florist's. If it is not her birthday and he brings no gift, the situation will be neutral, i.e., payoff 0. If it is not and he comes in bursting with roses, and obviously confused, he may be subjected to the Martini test, but he will emerge in a position of strong one-upness—which is worth 1. If it is her birthday and he has, clearly, remembered it, that is worth somewhat more, say 1.5. If he has forgotten it he is down like a stone, say, −10.

14. **The Hi-Fi** (J. D. Williams) The firm of Gunning & Kappler manufactures an amplifier. Its performance depends critically on the characteristics of one small, inaccessible condenser. This normally costs Gunning and Kappler $1, but they are set back a total of $10, on the average, if the original condenser is defective. There are some alternatives open to them. It is possible for them to buy a superior quality condenser, at $6, which is fully guaranteed; the manufacturer will make good the condenser and the costs incurred in getting the amplifier to operate. There is available also a condenser covered by an insurance policy which states, in effect "If it is our fault, we will bear the costs and you get your money back." This item costs $10. (This is a 3×2

game that Gunning & Kappler is playing against Nature whose options are to supply either a defective or a nondefective condenser.)

15. **The Huckster** (J. D. Williams) Merrill has a concession at the Yankee Stadium for the sale of sunglasses and umbrellas. The business places quite a strain on him, the weather being what it is. He has observed that he can sell about 500 umbrellas when it rains and about 100 when it shines; and in the latter case he also can dispose of about 1000 sunglasses. Umbrellas cost him 50 cents and sell for $1 (this is 1954); glasses cost 20 cents and sell for 50 cents. He is willing to invest $250 in the project. Everything that isn't sold is a total loss (the children play with them).

16. **The Coal Problem** (J. D. Williams) On a sultry summer afternoon, Hans' wandering mind alights upon the winter coal problem. It takes about 15 tons to heat his house during a normal winter, but he has observed extremes when as little as 10 tons and as much as 20 were used. He also recalls that the price per ton seems to fluctuate with the weather, being $10, $15, and $20 a ton during mild, normal, and severe winters. He can buy now, however, at $10 a ton. He considers three pure strategies, namely, to buy 10, 15, or 20 tons now and the rest, if any, later. He will be moving to California in the spring and he cannot take excess coal with him.

3

OPTIMAL RESPONSES
TO SPECIFIC STRATEGIES

The search for a player's optimal strategies for zero-sum games is initiated by an analysis of the situation where the opponent's strategy is known.

We shall now examine a player's options when he happens to know his opponent's strategy. Before doing so, however, it is necessary to caution against reading too much into this statement. Knowing the opponent's strategy is *not* tantamount to being able to predict the opponent's next decision. A strategy is merely a list that specifies the frequency with which each option should be played. Ideally speaking, once a player has decided on a strategy he will use some appropriate randomizing device to implement the strategy. For example, the strategy [.25, .5, .25] could make use of the spinner of Figure 1 or else the player could flip a coin twice and base his decision on the outcome as follows:

choose option 1 if two *heads* come up,

choose option 2 if a *head* and a *tail* come up,

choose option 3 if two *tails* come up.

In Bombing-sorties we considered the question of what Charlie should do if he observes that Ruth bluffs by placing the bomb on the support plane 1/4 of the time. Let us reconsider this question in a somewhat more formal manner. Using the notation and terminology of the previous chapter, Ruth's decision to bluff in this manner is tantamount to adopting the strategy [.75, .25] and Charlie's search for an appropriate response reduces to finding a strategy $[1 - q, q]$ such that the corresponding expected payoff (rate of successful missions) is as low as

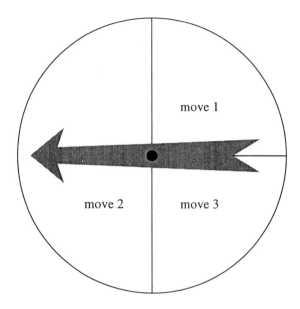

FIGURE 3.1. A randomizing spinner

possible. This search can now be made more methodical. The auxiliary diagram
that describes this situation is

	$1 - q$	q
.75	80%	100%
.25	90%	50%

and the expected payoff is

$$.75 \times (1 - q) \times 80\% + .75 \times q \times 100\% + .25 \times (1 - q) \times 90\% + .25 \times q \times 50\%$$
$$= .60(1 - q) + .75q + .225(1 - q) + .125q$$
$$= .60 - .60q + .75q + .225 - .225q + .125q$$
$$= .825 + .05q.$$

In other words, as a function of q, the expected payoff is $.825 + .05q$. Bearing in
mind that q is a probability and hence $0 \leq q \leq 1$, it follows that the expected
payoff is least when q is 0, when it assumes the value

$$.825 + .05 \times 0 = .825 = 82.5\%.$$

This means that Charlie's best response to Ruth's bluffing is to set $q = 0$ in
his general strategy $[1 - q, q]$. I.e. he should ignore the bluffing and consistently attack the bomber, just as was concluded in Chapter 1 on the basis of an
incomplete analysis.

Suppose that instead of bluffing 1/4 of the time Ruth decides that she will place the bomb on the support plane 1/2 the time, that is, suppose that Ruth adopts the strategy [.5, .5]. What is Charlie's best strategy then? Now the auxiliary diagram is

	$1-q$	q
.50	80%	100%
.50	90%	50%

and the expected payoff is

$$.50 \times (1-q) \times 80\% + .50 \times q \times 100\% + .50 \times (1-q) \times 90\% + .50 \times q \times 50\%$$
$$= .40(1-q) + .50q + .45(1-q) + .25q$$
$$= .40 - .40q + .50q + .45 - .45q + .25q$$
$$= .85 - .10q.$$

Since Charlie is trying to minimize payoffs it is clearly to his advantage to assign to q the largest possible value, namely 1. This means that under these circumstances Charlie's best strategy is $[1-1, 1] = [0, 1]$. In other words, if Ruth places the bomb (at random) on either plane with frequency 1/2, then it is to Charlie's advantage to attack the weaker support plane consistently.

It will prove convenient to introduce some new terms here. In the face of any specific strategy of Ruth's, that strategy of Charlie's that results in the *smallest* expected payoff to Ruth is called Charlie's *optimal counterstrategy*. Similarly, in the face of a specific strategy of Charlie's, that strategy of Ruth's that provides her with the *largest* expected payoff is called her *optimal counterstrategy*. The conclusion of the above discussion is that when Ruth employs the mixed strategy [.75, .25], Charlie's optimal counterstrategy is [1, 0] whereas when Ruth employs the strategy [.5, .5], Charlie's optimal counterstrategy is [0, 1]. That both of the counterstrategies are pure is no coincidence and we formulate the general principle as a theorem.

THEOREM 1. *If one player of a game employs a fixed strategy, then the opponent has an optimal counterstrategy that is pure.*

This theorem reduces the task of determining a player's optimal counterstrategy to his opponent's fixed strategy to a manageable number of computations. It should be pointed out that sometimes nonpure optimal counterstrategies are also available (see Example 3 below).

EXAMPLE 2. Find an optimal response for Charlie if it is known that Ruth is committed to the strategy [.2, .3, .5] in Rock-scissors-paper.

We know that Charlie has a pure optimal counterstrategy and so we compute the expected payoffs that correspond to the three pure strategies that are available to him. Making use of the auxiliary diagrams below, and ignoring the

	1	0	0
.2	0	1	−1
.3	−1	0	1
.5	1	−1	0

	0	1	0
.2	0	1	−1
.3	−1	0	1
.5	1	−1	0

	0	0	1
.2	0	1	−1
.3	−1	0	1
.5	1	−1	0

columns that are used with 0 frequencies, we get:

for $[1, 0, 0]$:

$$.2 \times 1 \times 0 + .3 \times 1 \times (-1) + .5 \times 1 \times 1 = 0 - .3 + .5 = .2,$$

for $[0, 1, 0]$:

$$.2 \times 1 \times 1 + .3 \times 1 \times 0 + .5 \times 1 \times (-1) = .2 + 0 - .5 = -.3,$$

for $[0, 0, 1]$:

$$.2 \times 1 \times (-1) + .3 \times 1 \times 1 + .5 \times 1 \times 0 = -.2 + .3 + 0 = .1.$$

Since the expected payoff denotes Ruth's winnings, Charlie should opt for the minimum payoff of $-.3$ by choosing the pure strategy $[0, 1, 0]$, i.e., by consistently displaying the *scissors*. Note that this agrees with our intuition. After all, Ruth's commitment to the strategy $[.2, .3, .5]$ favors the *paper* option. It therefore makes sense that Charlie should capitalize on this information by favoring the *scissors*, since they win over Ruth's favorite *paper*. What is perhaps surprising, is the conclusion that under these circumstances Charlie should not only favor *scissors* but in fact use it exclusively. Such, however, is the implication of Theorem 1 above.

EXAMPLE 3. Find an optimal counterstrategy for Ruth if it is known that Charlie is committed to the strategy $[.5, .4, .1]$ in the abstract zero-sum game

1	1	1
2	0	3
−4	5	10
3	−1	2

The auxiliary diagrams that correspond to Ruth's four pure options are

	.5	.4	.1
1	1	1	1
0	2	0	3
0	−4	5	10
0	3	−1	2

	.5	.4	.1
0	1	1	1
1	2	0	3
0	−4	5	10
0	3	−1	2

	.5	.4	.1
0	1	1	1
0	2	0	3
1	−4	5	10
0	3	−1	2

	.5	.4	.1
0	1	1	1
0	2	0	3
0	−4	5	10
1	3	−1	2

and the corresponding expected payoffs are:

for $[1, 0, 0, 0]$:

$$1 \times .5 \times 1 + 1 \times .4 \times 1 + 1 \times .1 \times 1 = .5 + .4 + .1 = 1,$$

for $[0, 1, 0, 0]$:

$$1 \times .5 \times 2 + 1 \times .4 \times 0 + 1 \times .1 \times 3 = 1 + 0 + .3 = 1.3,$$

for $[0, 0, 1, 0]$:

$$1 \times .5 \times (-4) + 1 \times .4 \times 5 + 1 \times .1 \times 10 = -2 + 2 + 1 = 1,$$

for $[0, 0, 0, 1]$:

$$1 \times .5 \times 3 + 1 \times .4 \times (-1) + 1 \times .1 \times 2 = 1.5 - .4 + .2 = 1.3.$$

Since Ruth wishes to maximize the expected payoff she will aim for the 1.3. She can use either of the pure strategies $[0, 1, 0, 0]$ or $[0, 0, 0, 1]$ to guarantee this expectation and optimize her response. In fact, any mix of these two strategies will also provide the same guarantee. Thus, Ruth's mixed strategy $[0, .3, 0, .7]$ yields the expected payoff

	.5	.4	.1
0	1	1	1
.3	2	0	3
0	−4	5	10
.7	3	−1	2

$$.3 \times .5 \times 2 + .3 \times .4 \times 0 + .3 \times .1 \times 3 + .7 \times .5 \times 3 + .7 \times .4 \times (-1) + .7 \times .1 \times 2$$
$$= .3 + 0 + .09 + 1.05 - .28 + .14 = 1.3.$$

The principle in question here is made explicit in the following theorem.

THEOREM 4. *In any zero-sum game, if one player employs a fixed strategy, then any mixture of the opponent's pure optimal counterstrategies is itself a mixed optimal counterstrategy.*

Proofs of 2 × 2 Cases*

THEOREM 5. *In any 2 × 2 zero-sum game, if one player employs a fixed strategy, then the opponent has an optimal counterstrategy that is pure.*

PROOF. Suppose $E(p, q)$ is the expected payoff when Ruth and Charlie employ the strategies $[1 - p, p]$ and $[1 - q, q]$ respectively in the game

$$G = \begin{array}{|c|c|} \hline a & b \\ \hline c & d \\ \hline \end{array}$$

Then

$$E(p, q) = (1 - p)(1 - q)a + (1 - p)qb + p(1 - q)c + pqd$$
$$= p(-a + aq + c - bq - cq + dq) + (a - aq + bq).$$

If Charlie employs a *fixed* strategy $[1 - q, q]$ then the quantities a, b, c, d, q are all fixed and so $E(p, q)$, as a function of p, $0 \leq p \leq 1$, assumes its minimum value at

$$p = \begin{cases} 0 & \text{if } -a + aq + c - bq - cq + dq \geq 0 \\ 1 & \text{if } -a + aq + c - bq - cq + dq \leq 0. \end{cases}$$

In either case, Ruth has a pure optimal counterstrategy. The proof that Charlie has pure optimal counterstrategies is relegated to Exercise 21.

Alternately,

$$E(p, q) = (1 - p)((1 - q)a + qb) + p((1 - q)c + qd)$$
$$= (1 - p)E(0, q) + pE(1, q).$$

Hence $E(p, q)$, being a weighted average of $E(0, q)$ and $E(1, q)$, must lie between them. Consequently Ruth's optimal response to Charlie's $[1 - q, q]$ can be obtained by setting p to be either 0 if $E(0, q) \geq E(1, q)$ or 1 if $E(1, q) \geq E(0, q)$. In other words, Ruth has a pure optimal counterstrategy. A similar proof works for Charlie (see Exercise 22).   q.e.d.

Chapter Summary

We considered the situation where one player has fixed on a specific strategy. Under these circumstances, the opponent can optimize his response with a pure counterstrategy. It may happen that several pure strategies will serve to optimize the opponent's response, and in that case any mixture of these optimal pure responses is itself a mixed optimal counterstrategy.

Chapter Terms

Optimal counterstrategy   27

EXERCISES 3

In Exercises 1–18 $\mathbf{R}$ denotes a fixed strategy of Ruth's and $\mathbf{C}$ denotes a fixed strategy of Charlie's for the given game $\mathbf{G}$. In each case find an optimal counterstrategy for the opponent.

1. $\mathbf{R} = [.1, .9]$, $\mathbf{G} =$

−1	3
4	−2

2. $\mathbf{R} = [.8, .2]$, $\mathbf{G} =$

−1	3
4	−2

3. $\mathbf{R} = [.5, .5]$, $\mathbf{G} =$

−1	3
4	−2

4. $\mathbf{C} = [.1, .9]$, $\mathbf{G} =$

−1	3
4	−2

5. $\mathbf{C} = [.8, .2]$, $\mathbf{G} =$

−1	3
4	−2

6. $\mathbf{C} = [.5, .5]$, $\mathbf{G} =$

−1	3
4	−2

7. $\mathbf{R} = [.1, .7, .2]$, $\mathbf{G} =$

1	0	−2	3
−3	4	2	−4
0	−1	0	1

8. $\mathbf{R} = [.5, .5, 0]$, $\mathbf{G} =$

1	0	−2	3
−3	4	2	−4
0	−1	0	1

9. $\mathbf{R} = [.8, .1, .1]$, $\mathbf{G} =$

1	0	−2	3
−3	4	2	−4
0	−1	0	1

1	0	−2	3
−3	4	2	−4
0	−1	0	1

10. $\mathbf{C} = [.1, .7, .1, .1]$, $\mathbf{G} =$

1	0	−2	3
−3	4	2	−4
0	−1	0	1

11. $\mathbf{C} = [.5, 0, .5, 0]$, $\mathbf{G} =$

1	0	−2	3
−3	4	2	−4
0	−1	0	1

12. $\mathbf{C} = [.8, .1, 0, .1]$, $\mathbf{G} =$

1	−3	2
0	4	−4
−2	0	2
3	−3	−1
−3	5	1

13. $\mathbf{R} = [.2, .3, .2, .2, .1]$, $\mathbf{G} =$

1	−3	2
0	4	−4
−2	0	2
3	−3	−1
−3	5	1

14. $\mathbf{R} = [0, .4, 0, .6, 0]$, $\mathbf{G} =$

1	−3	2
0	4	−4
−2	0	2
3	−3	−1
−3	5	1

15. $\mathbf{R} = [.4, 0, .3, 0, .3]$, $\mathbf{G} =$

16. $\mathbf{C} = [.1, .7, .2]$, $\mathbf{G} =$

1	−3	2
0	4	−4
−2	0	2
3	−3	−1
−3	5	1

17. $\mathbf{C} = [.7, .2, .1]$, $\mathbf{G} =$

1	−3	2
0	4	−4
−2	0	2
3	−3	−1
−3	5	1

18. $\mathbf{C} = [.2, .1, .7]$, $\mathbf{G} =$

1	−3	2
0	4	−4
−2	0	2
3	−3	−1
−3	5	1

19*. Prove Theorem 1.

20*. Prove Theorem 4.

21*. Using the first method in the proof of Theorem 5, prove that if Ruth employs a fixed strategy for the 2×2 game G then Charlie has an optimal counterstrategy that is pure.

22*. Using the second method in the proof of Theorem 5, prove that if Ruth employs a fixed strategy for the 2×2 game G then Charlie has an optimal counterstrategy that is pure.

THE MAXIMIN STRATEGY

A good strategy for Ruth is defined and proposed.

We shall now devise and justify a good strategy for Ruth when she plays the general 2×2 zero-sum game

a	b
c	d

Because of Theorem 3.1 above, Ruth knows that for any strategy $[1 - p, p]$ that she employs she can expect Charlie to pursue a counterstrategy that returns to her the lesser of the payoffs coming from the two *pure* strategies that are available for him. Let $r_1(p)$ and $r_2(p)$ be the expected payoffs that come from Charlie's pure strategies $[1, 0]$ and $[0, 1]$ respectively. Here the auxiliary diagrams are

$$
\begin{array}{cc}
 & \begin{array}{cc} 1 & 0 \end{array} \\
\begin{array}{c} 1 - p \\ p \end{array} & \begin{array}{|c|c|} \hline a & b \\ \hline c & d \\ \hline \end{array} \\
 & r_1(p)
\end{array}
\qquad
\begin{array}{cc}
 & \begin{array}{cc} 0 & 1 \end{array} \\
\begin{array}{c} 1 - p \\ p \end{array} & \begin{array}{|c|c|} \hline a & b \\ \hline c & d \\ \hline \end{array} \\
 & r_2(p)
\end{array}
$$

and we compute

$$r_1(p) = (1 - p) \times 1 \times a + p \times 1 \times c = a(1 - p) + cp = (c - a)p + a,$$
$$r_2(p) = (1 - p) \times 1 \times b + p \times 1 \times d = b(1 - p) + dp = (d - b)p + b.$$

If $E_R(p)$ denotes the expected payoff selected from $r_1(p)$ and $r_2(p)$ by Charlie, then, since Charlie can be relied on to lower this payoff as much as he can,

$$E_R(p) = \text{the lesser of } \{r_1(p), r_2(p)\}. \tag{1}$$

This completely determines Ruth's expected payoff, $E_R(p)$, as a function of p, i.e., as a function of her strategy. We shall use the graph of this function in order to suggest a strategy for Ruth.

Since the independent variable p appears in the expressions of $r_1(p)$ and $r_2(p)$ with degree at most 1, the graphs of these functions are straight lines. Inasmuch as p denotes a probability, we have $0 \le p \le 1$, and so these graphs consist of line segments that lie over the interval $[0, 1]$ on the p-axis. More specifically, since

$$r_1(0) = a \times (1 - 0) + c \times 0 = a$$
$$r_1(1) = a \times (1 - 1) + c \times 1 = c,$$

it follows that the graph of $r_1(p)$ is the line segment joining the points $(0, a)$ and $(1, c)$. Similarly, since

$$r_2(0) = b \times (1 - 0) + d \times 0 = b$$
$$r_2(1) = b \times (1 - 1) + d \times 1 = d,$$

it follows that the graph of $r_2(p)$ consists of the straight line segment joining $(0, b)$ and $(1, d)$. It will be seen that this makes the sketching of the graphs of $r_1(p)$ and $r_2(p)$ a very easy matter (it is for this reason that we chose to denote Ruth's general strategy by $[1 - p, p]$ rather than $[p, 1 - p]$, since the latter would have reversed the graphs, thereby adding an unnecessary complication). The graph of $E_R(p)$ is then also easily derived according to the following observation:

The graph of $E_R(p)$ consists of that line which, for every permissible value of p, contains the lower of the two points $(p, r_1(p))$ and $(p, r_2(p))$.

The subsequent examples will demonstrate that much useful information can be read from this graph.

EXAMPLE 1. For Penny-matching, $a = d = 1$, $b = c = -1$, and the graph of

	Head	Tail
Head	1	−1
Tail	−1	1

$E_R(p)$ is the broken heavy line in Figure 1.

Since this graph coincides with that of $r_2(p)$ for $0 \le p < .5$, it follows that if Ruth employs a strategy $[1 - p, p]$ with $p < .5$, then Charlie should respond with the pure strategy $[0, 1]$. In other words, if Ruth favors *heads*, then Charlie should respond by showing *tails all the time*, a conclusion that is intuitively plausible. On the other hand, if Ruth favors *tails* and employs a strategy with $p > .5$, then, since the graph of $E_R(p)$ now coincides with that of $r_1(p)$, it follows that Charlie should employ the pure strategy $[1, 0]$ and play *heads* all the time. When $p = .5$, the graph of $E_R(p)$ coincides with both the graphs of $r_1(p)$ and $r_2(p)$ and so it does not matter which strategy is employed by Charlie. We summarize this by

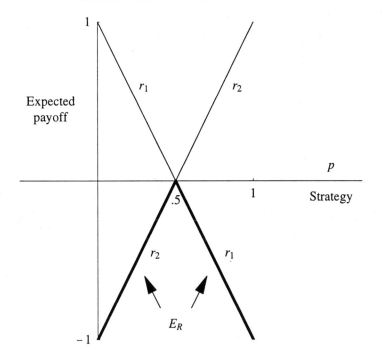

FIGURE 4.1. A graph of Ruth's expectation.

saying that

$[0, 1]$ is an optimal counterstrategy for Charlie when $p \leq .5$,

$[1, 0]$ is an optimal counterstrategy for Charlie when $p \geq .5$.

Moreover, since the highest point on the graph of $E_R(p)$ is the one that corresponds to $p = .5$, this figure tells us that Ruth had best employ the strategy $[1 - .5, .5] = [.5, .5]$ since that is the one that guarantees her the largest expected payoff, namely, 0. This conclusion, too, fits in well with our intuition.

EXAMPLE 2. For the Bombing sorties game

80%	100%
90%	50%

we have $a = 80\% = .8$, $b = 100\% = 1$, $c = 90\% = .9$, $d = 50\% = .5$,

$$r_1(p) = (.9 - .8)p + .8 = .1p + .8,$$
$$r_2(p) = (.5 - 1)p + 1 = -.5p + 1,$$

and so the graph of $E_R(p)$ is the heavy broken line of Figure 2.

As the graph of $E_R(p)$ coincides with that of $r_1(p)$ for small values of p, it follows that when Ruth employs the strategy $[1 - p, p]$ for small values of p,

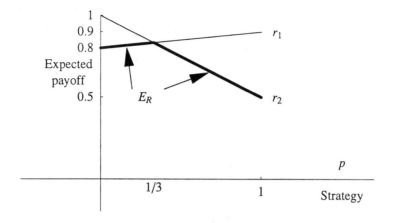

FIGURE 4.2. A graph of Ruth's expectation.

Charlie should counter with the pure strategy $[1,0]$ and when p is close to 1, Charlie should respond with $[0,1]$. The cutoff point is of course the value of p that lies directly below the point of intersection of the graphs of $r_1(p)$ and $r_2(p)$. This important value is found by solving the equation

$$r_1(p) = r_2(p)$$

or

$$.1p + .8 = -.5p + 1$$
$$.6p = .2$$
$$p = \frac{.2}{.6} = \frac{2}{6} = \frac{1}{3}.$$

Hence,

$[1,0]$ is an optimal counterstrategy for Charlie when $p \leq 1/3$,

$[0,1]$ is an optimal counterstrategy for Charlie when $p \geq 1/3$.

In other words, as long Ruth places the bomb on the support plane less than $1/3$ of the time, Charlie should persist in attacking the bomber. Once the bomb is placed on the support plane with a frequency greater than $1/3$, Charlie should switch to attacking this weaker plane consistently. When $p = 1/3$ all of Charlie's strategies will yield the same expected payoff.

Since the intersection of the graphs of $r_1(p)$ and $r_2(p)$ also happens to be the highest point on the graph of $E_R(p)$, it also corresponds to Ruth's wisest choice. By employing the strategy

$$\left[1 - \frac{1}{3}, \frac{1}{3}\right] = \left[\frac{2}{3}, \frac{1}{3}\right]$$

Ruth obtains the largest possible expected payoff she can guarantee. The exact value of this largest expected payoff can be computed by substituting $p = 1/3$ into either $r_1(p)$ or $r_2(p)$:

$$r_1\left(\frac{1}{3}\right) = .1 \times \frac{1}{3} + .8 = .8333\ldots = 83.\bar{3}\%$$

or

$$r_2\left(\frac{1}{3}\right) = -.5 \times \frac{1}{3} + 1 = .8333\ldots = 83.\bar{3}\%.$$

It is clear from the foregoing examples that the highest point of the graph of $E_R(p)$ is of special strategic significance. Unfortunately, this graph may have more than one highest point (see Example 6 below) and so some care must be exercised in stating following definition/theorem.

DEFINITION/THEOREM 3. *If (x, y) is any highest point on the graph of $E_R(p)$ then*

$$[1 - x, x] \text{ is a maximin strategy for Ruth, and}$$

$$y \text{ is Ruth's maximin expectation.}$$

If Ruth employs the maximin strategy $[1 - x, x]$ then she can expect to win, on the average, at least y on each play.

The reason for this nomenclature is that every point on this graph is the lesser (minimum) of the two choices available to Charlie, and Ruth, in choosing the highest point on this graph is *maximizing* Charlie's *minimal* return. Speaking informally, the maximum strategy and expectation are given by *the high point on the low curve*. Thus, the maximin strategy in Penny-matching is $[1/2, 1/2]$ and Ruth's maximin expectation is 0. Similarly, the maximin strategy in Bombing-sorties is $[2/3, 1/3]$ and Ruth's maximin expectation is $83.\bar{3}\%$. The next example demonstrates that the maximin strategy need not correspond to the intersection of $r_1(p)$ and $r_2(p)$.

EXAMPLE 4. In the abstract game

0	−1
2	3

$a = 0$, $b = -1$, $c = 2$, $d = 3$, and

$$r_1(p) = (2 - 0)p + 0 = 2p,$$
$$r_2(p) = (3 - (-1))p + (-1) = 4p - 1.$$

The graph of $E_R(p)$ is the broken heavy line of Figure 3.

Note that in this figure the unit on the expected payoff axis has a different length from the unit on the Strategy (or p) axis. This is convenient because in most of the subsequent games the payoffs will be integers, sometimes fairly large ones. The resulting distortion will have no effect on the validity of the

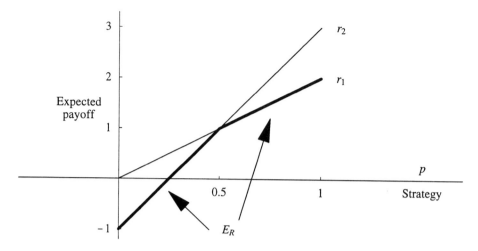

FIGURE 4.3. A graph of Ruth's expectation.

conclusions. To find the intersection of the graphs of $r_1(p)$ and $r_2(p)$ we set $r_1(p) = r_2(p)$ and solve for p:

$$2p = 4p - 1$$
$$-2p = -1$$
$$p = \frac{-1}{-2} = \frac{1}{2}.$$

Thus, Charlie should employ the pure strategy $[1, 0]$ or $[0, 1]$ according as Ruth favors her second or her first option.

In this case the highest point on the graph of $E_R(p)$ is *not* the intersection of the graphs of $r_1(p)$ and $r_2(p)$. Rather, it is the point $(1, 2)$. Hence, the maximin strategy of Ruth is the pure strategy $[1 - 1, 1] = [0, 1]$. In retrospect, this makes good sense. Since every entry of the second row of this game is larger than the entry above it, Ruth only stands to lose by selecting the first row, no matter what Charlie does. Thus, Ruth should select the second row consistently; i.e., she should use the pure strategy $[0, 1]$.

The maximin expectation, the y-coordinate of the highest point $(1, 2)$ is 2. This example underscores the fact that this maximin expectation constitutes a floor, or a lower bound, for Ruth's expectations. If she employs the maximin strategy of $[0, 1]$, then she can expect to win at least 2 each time the game is played. However, should Charlie play foolishly (i.e., should he make occasional use of his second column), then Ruth might win more.

EXAMPLE 5. In the abstract game

3	1
−1	−2

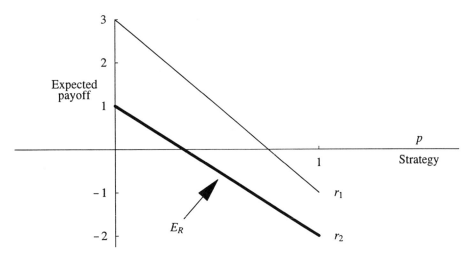

FIGURE 4.4. A graph of Ruth's expectation.

$a = 3$, $b = 1$, $c = -1$, $d = -2$ and

$$r_1(p) = (-1 - 3)p + 3 = -4p + 3,$$
$$r_2(p) = (-2 - 1)p + 1 = -3p + 1.$$

In this case, as is clear from Figure 4, the graph of $E_R(p)$ coincides with the graph of $r_2(p)$ for all $0 \leq p \leq 1$. This means that regardless of Ruth's specific strategy, Charlie should always select his second column. Actually, we didn't really need the graph to see this. Since every entry of the second column is less than the entry on its left, Charlie only stands to lose by ever employing the first column.

Ruth's maximin strategy, the one that yields the largest guaranteed expected payoff, comes from the left endpoint of the graph of $E_R(p)$, namely, the one above $p = 0$. Thus, it is the pure strategy $[1 - 0, 0] = [1, 0]$. Inasmuch as this strategy dictates that Ruth should select the first row exclusively, it follows that this strategy guarantees Ruth an expected payoff of 1 (this being the least entry in the first row).

In all of the above examples Ruth's maximin strategy turned out to be unique. This need not always be the case. For example, in the game

0	0
0	0

it clearly does not matter what either player does, and consequently any strategy of Ruth's is a maximin strategy guaranteeing the value 0. A more interesting example is offered below.

EXAMPLE 6. The abstract game

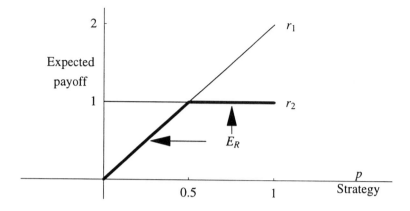

FIGURE 4.5. A graph of Ruth's expectation.

0	1
2	1

has $a = 0$, $c = 2$, $b = d = 1$ and

$$r_1(p) = (2 - 0)p + 0 = 2p,$$
$$r_2(p) = (1 - 1)p + 1 = 1.$$

The graph of $E_R(p)$ is the heavy line of Figure 5. Rather than a peak, this graph has a plateau consisting of the heavy line segment to the right of the intersection of the graphs of $r_1(p)$ and $r_2(p)$. This point of intersection is again obtained by setting $r_1(p) = r_2(p)$:

$$2p = 1 \quad \text{or} \quad p = .5.$$

Hence any strategy $[1-p, p]$ with $p \geq .5$ is a maximin strategy for Ruth, including the mixed strategy $[.5, .5]$ and the pure strategy $[0, 1]$. Any of these strategies will guarantee Ruth an expected payoff of 1 (the height of the plateau).

In conclusion, the maximin strategy is good in the sense that it guarantees Ruth a certain expected payoff and, moreover, this is the best expected payoff that can be guaranteed. Nonetheless, it is natural to ask now whether the maximin strategy is the best strategy? The answer to this question, of course, depends on one's yardstick. If the yardstick is that of absolute guarantees, then the maximin strategy is definitely not the best. For example, suppose the game

5	−3
2	4

is played 10 times and Ruth employs the maximin strategy of $[.2, .8]$. If she happens to select the first row on the first, fifth, and ninth plays only, and if Charlie happens to select the second column on those same plays only, then

Ruth ends up with a total win of $7 \cdot 2 + 3 \cdot (-3) = 5$, which is much less than the total of 20 she would have been sure to win had she always selected the second row. What the maximin strategy is best at, is guaranteeing the expected payoff, which in the above example equals $10(.2 \cdot 5 + .8 \cdot 2) = 26$ which is considerably better than the aforementioned absolute guarantee of 20.

Von Neumann and Morgenstern were aware of the relativistic value of the maximin strategy. They viewed it as defensive strategy since it protected one's expected payoff. In their words:

> All this may be summed up by saying that while our good strategies are perfect from the defensive point of view, they will (in general) not get the maximum out of the opponent's (possible) mistakes,—i.e. they are not calculated for the offensive.
>
> It should be remembered, however, that our deductions of 17.8 are nevertheless cogent; i.e. a theory of the offensive, in this sense, is not possible without essentially new ideas. The reader who is reluctant to accept this, ought to visualize the situation in Matching Pennies or Stone, Paper, Scissors once more; the extreme simplicity of these two games makes the decisive points particularly clear.

Chapter Summary

In any 2×2 zero-sum game, the function $E_R(p)$ denotes the expected payoff that Ruth can look forward to when she employs the strategy $[1 - p, p]$. The graph of this function is easily drawn. If (x, y) are the coordinates of any highest point on this graph then $[1 - x, x]$ is the recommended maximin strategy for Ruth, and $y = E_R(x)$ is the corresponding payoff. Of all the mixed strategies available to her, this maximin strategy $[1 - x, x]$ provides Ruth with the best guarantee on her *expected* payoff.

Chapter Terms

$E_R(p)$	35	Maximin expectation	39
Maximin strategy	39		

EXERCISES 4

For each of the games in Exercises 1–15,
  a) draw the graph of $E_R(p)$,
  b) determine a maximin strategy,
  c) find Ruth's maximin expectation,
  d*) specify for which values of p Charlie's optimal counterstrategy is $[1, 0]$ and for which it is $[0, 1]$.

1.
1	3
2	1

2.
3	2
0	2

3.
2	1
3	1

4.
2	1
1	3

5.
2	1
4	6

6.
1	-3
-2	1

7.

1	3
-2	1

8.

0	-1
0	1

9.

2	4
0	-1

10.

1	1
2	-1

11.

1	1
-3	1

12.

-2	1
0	3

13.

5	-1
-4	3

14.

-1	3
6	-4

15.

8	2
2	-8

THE MINIMAX STRATEGY

A good strategy for Charlie is defined and proposed.

We now turn to Charlie and devise for him too a good strategy. The graph of Charlie's expected payoff in 2×2 zero-sum games is obtained in much the same way as Ruth's graph, with some important differences, and similar conclusions can be drawn.

Because of Theorem 1 of Chapter 3, Charlie knows that for any strategy $[1 - q, q]$ that he employs in the general 2×2 game

a	b
c	d

he can expect Ruth to pursue a counterstrategy that returns to her the larger of the payoffs coming from the two *pure* strategies that are available to her. Let $c_1(q)$ and $c_2(q)$ denote the expected payoffs that come from Ruth's pure strategies $[1, 0]$ and $[0, 1]$ respectively. Here the auxiliary diagrams are

$$
\begin{array}{cc}
 & 1-q \quad q \\
\begin{array}{c}1\\[1.2em]0\end{array} &
\begin{array}{|c|c|}
\hline
a & b \\
\hline
c & d \\
\hline
\end{array}
\end{array}
\qquad
\begin{array}{cc}
 & 1-q \quad q \\
\begin{array}{c}0\\[1.2em]1\end{array} &
\begin{array}{|c|c|}
\hline
a & b \\
\hline
c & d \\
\hline
\end{array}
\end{array}
$$
$$
\qquad\quad c_1(q) \qquad\qquad\qquad\quad c_2(q)
$$

and the expected payoffs are

$$
c_1(q) = 1 \times (1 - q) \times a + 1 \times q \times b = a(1 - q) + bq = (b - a)q + a,
$$
$$
c_2(q) = 1 \times (1 - q) \times c + 1 \times q \times d = c(1 - q) + dq = (d - c)q + c.
$$

If $E_C(q)$ denotes the expected payoff selected from $c_1(q)$ and $c_2(q)$ by Ruth, then, since Ruth aims to maximize her gains,

$$E_C(q) = \text{the larger of } \{c_1(q), c_2(q)\} \tag{1}$$

This completely determines the expected payoff $E_C(q)$ as a function of q, i.e., as a function of Charlie's strategy. We shall use the graph of this function in order to suggest a good strategy for Charlie.

Inasmuch as the independent variable q appears in the expressions of $c_1(q)$ and $c_2(q)$ with degree at most 1, the graphs of these functions are straight lines. Since q denotes a probability, we have $0 \le q \le 1$, and so these graphs consist of line segments that lie over the interval $[0, 1]$ on the q-axis. More specifically, since

$$c_1(0) = a \times (1 - 0) + b \times 0 = a$$
$$c_1(1) = a \times (1 - 1) + b \times 1 = b,$$

it follows that the graph of $c_1(q)$ is the line segment joining the points $(0, a)$ and $(1, b)$. Similarly, since

$$c_2(0) = c \times (1 - 0) + d \times 0 = c$$
$$c_2(1) = c \times (1 - 1) + d \times 1 = d,$$

it follows that the graph of $c_2(q)$ consists of the straight line segment joining $(0, c)$ and $(1, d)$. Again this makes the sketching of the graphs of $c_1(q)$ and $c_2(q)$ a very easy matter. The graph of $E_C(q)$ is then also easily derived according to the following observation:

The graph of $E_C(q)$ consists of that line which, for every permissible value of q, contains the higher of the two points $(q, c_1(q))$ and $(q, c_2(q))$.

We shall now reexamine each of the games of the previous chapter from Charlie's point of view.

EXAMPLE 1. For the Penny-matching game $a = d = 1$, $b = c = -1$, and

1	-1
-1	1

the graph of $E_C(q)$ is the broken heavy line in Figure 1.

Since this graph coincides with that of $c_1(q)$ for $0 \le q < .5$, it follows that if Charlie employs a strategy $[1 - q, q]$ with $q < .5$, then Ruth should respond with the pure strategy $[1, 0]$. In other words, if Charlie favors *heads*, then Ruth should respond by showing *heads* **all the time**, a conclusion that is intuitively plausible. On the other hand, if Charlie favors *tails* by employing a strategy with $q > .5$, then, since the graph of $E_C(q)$ now coincides with that $c_2(q)$, it follows that Ruth should follow the pure strategy $[0, 1]$ and play *tails* all the

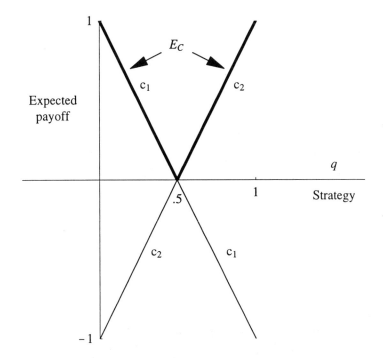

FIGURE 5.1. A graph of Charlie's expectation.

time. When $q = .5$ all of Ruth's strategies will yield the same expected payoff. This is summarized as

[1, 0] is an optimal counterstrategy for Ruth when $q \leq .5$,

[0, 1] is an optimal counterstrategy for Ruth when $q \geq .5$.

Moreover, since the lowest point on the graph of $E_C(q)$ is the one above $q = .5$, this figure tells us that Charlie had best employ the strategy $[1 - .5, .5] = [.5, .5]$ since that is the one that guarantees the lowest possible expected payoff for Ruth, namely, 0. This conclusion, too, fits in well with our intuition.

EXAMPLE 2. For the Bombing-sorties game

80%	100%
90%	50%

we have $a = 80\% = .8$, $b = 100\% = 1$, $c = 90\% = .9$, $d = 50\% = .5$,

$$c_1(q) = (1 - .8)q + .8 = .2q + .8,$$
$$c_2(q) = (.5 - .9)q + .9 = -.4q + .9,$$

and so the graph of $E_C(q)$ is the heavy broken line of Figure 2.

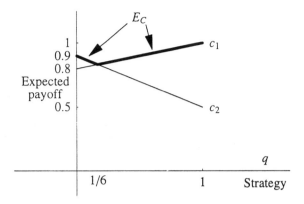

FIGURE 5.2. A graph of Charlie's expectation.

Since the graph of $E_C(q)$ coincides with that of $c_2(q)$ for small values of q, it follows that when Charlie employs the strategy $[1 - q, q]$ for small values of q, Ruth should respond with the pure counterstrategy $[0, 1]$; when q is close to 1, Ruth should respond with $[1, 0]$. In other words, if Charlie is observed to hardly ever attack the support plane (i.e., if q is small), Ruth should always place the bomb on the support plane; if Charlie is observed to mostly attack to support plane (i.e., if q is close to 1), Ruth should always put the bomb on the bomber. The cutoff point is of course the value of q that lies directly below the point of intersection of the graphs of $c_1(q)$ and $c_2(q)$. This value is found by solving the equation

$$c_1(q) = c_2(q)$$

or

$$.2q + .8 = -.4q + .9$$
$$.6q = .1$$
$$q = \frac{.1}{.6} = \frac{1}{6}.$$

Hence,

$[0, 1]$ is an optimal counterstrategy for Ruth when $q \leq 1/6$,

$[1, 0]$ is an optimal counterstrategy for Ruth when $q \geq 1/6$.

In other words, as long as Charlie attacks the support plane no more than $1/6$ of the time, Ruth should persist in placing the bomb on this lighter plane. Once the support plane is attacked with a frequency greater than $1/6$, Ruth should switch to placing the bomb on the bomber consistently. When $q = 1/6$ all of Ruth's strategies yield the same expected payoff.

Since the intersection of the graphs of $c_1(q)$ and $c_2(q)$ also happens to be the lowest point on the graph of $E_C(q)$, it also corresponds to Charlie's wisest strategy. By employing the strategy

$$\left[1 - \frac{1}{6}, \frac{1}{6}\right] = \left[\frac{5}{6}, \frac{1}{6}\right]$$

Charlie lowers the expected payoff as much as he can. The exact value of this lowest expected payoff can be computed by substituting $q = 1/6$ into either $c_1(q)$ or $c_2(q)$:

$$c_1\left(\frac{1}{6}\right) = .2 \times \frac{1}{6} + .8 = .8333\ldots = 83.\overline{3}\%$$

or

$$c_2\left(\frac{1}{6}\right) = -.4 \times \frac{1}{6} + .9 = .8333\ldots = 83.\overline{3}\%.$$

It is clear from the foregoing examples that the lowest point of the graph of $E_C(q)$ is of special strategic significance. As was the case with the highest points of the graph of $E_R(p)$, the graph of $E_C(q)$ may have more than one of these lowest points. We now state a definition that is the analog of the main concepts of the previous chapter.

DEFINITION/THEOREM 3. *If (x, y) is any lowest point on the graph of $E_C(q)$ then*

$$[1 - x, x] \textit{ is a minimax strategy for Charlie, and}$$
$$y \textit{ is Ruth's minimax expectation.}$$

If Charlie employs the minimax strategy $[1 - x, x]$ then he can expect to hold Ruth's average winnings to no more that y.

The reason for this nomenclature is that every point on this graph is the larger (maximum) of the two choices available to Ruth, and Charlie, in choosing the lowest point on this graph is **minimizing** Ruth's **maximum** return. Speaking informally, the minimax strategy and expectation are given by *the low point on the high curve.* Thus, the minimax strategy in Penny-matching is $[1/2, 1/2]$ and Ruth's minimax expectation is 0. Similarly, the minimax strategy in Bombing sorties is $[5/6, 1/6]$ and Ruth's minimax expectation is $83.\overline{3}\%$. The next example demonstrates that if Ruth is not careful, then even if Charlie is employing a minimax strategy she may win less than the minimax expectation.

EXAMPLE 4. In the abstract game

0	-1
2	3

$a = 0$, $b = -1$, $c = 2$, $d = 3$, and

$$c_1(q) = (-1 - 0)q + 0 = -q,$$
$$c_2(q) = (3 - 2)q + 2 = q + 2.$$

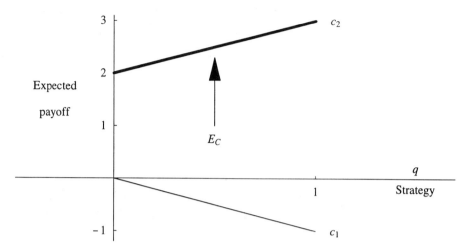

FIGURE 5.3. A graph of Charlie's expectation.

It is clear from Figure 3 that here the graph of $E_C(q)$ coincides with that of $c_2(q)$. Since the low point on this graph corresponds to $q = 0$, it follows that Charlie's minimax strategy is the pure strategy $[1, 0]$. By pursuing this strategy all Charlie can accomplish is to hold Ruth's winnings down to no more than 2 per play, but that is the best he can do under the circumstances. This game is patently unfair to Charlie.

The minimax expectation, the y-coordinate of $(0, 2)$ which is the lowest point of the graph of $E_C(q)$, is 2. This game underscores the fact that this minimax expectation constitutes a ceiling, or an upper bound, on Ruth's expectation. If Charlie employs the minimax strategy $[1, 0]$ and Ruth is foolish enough to occasionally select the first row, the she will average less than 2 per play. On the other hand, if she plays wisely, then she can win 2 (which equals the minimax expectation) in each play.

EXAMPLE 5. In the abstract game

3	1
−1	−2

$a = 3$, $b = 1$, $c = -1$, $d = -2$ and

$$c_1(q) = (1 - 3)q + 3 = -2q + 3,$$
$$c_2(q) = (-2 + 1)q - 1 = -q - 1.$$

In this case, as is clear from Figure 4, the graph of $E_C(q)$ coincides with the graph of $c_1(q)$. This means that regardless of Charlie's specific strategy, Ruth should always employ her first option. Again, we didn't really need the graph to see this. Since every entry of the first row is greater than the entry below it, Ruth only stands to lose by ever employing the second row.

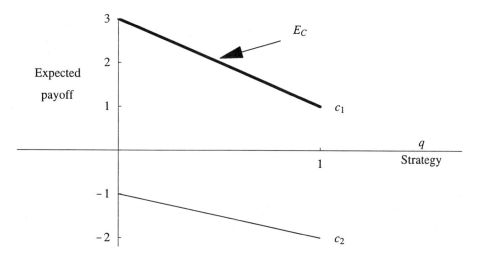

FIGURE 5.4. A graph of Charlie's expectation.

Charlie's minimax strategy, the one that yields the smallest expected payoff to Ruth, comes from the right endpoint of the graph of $E_C(q)$, namely, the one above $q = 1$. It is the pure strategy $[1 - 1, 1] = [0, 1]$. Since the largest entry of the corresponding second column is 1, this will result in Charlie's holding Ruth's winnings down to at most 1 per play. As was the case for Ruth's maximin strategy, Charlie's minimax strategy need not be unique.

Chapter Summary

In any 2×2 zero-sum game, the function $E_C(q)$ denotes the expected payoff that Ruth can look forward to when Charlie employs the strategy $[1 - q, q]$. The graph of this function is easily drawn. If (x, y) are the coordinates of any lowest point of this graph, then $[1 - x, x]$ is the recommended minimax strategy for Ruth, and $y = E_C[x]$ is the minimax expectation. This minimax expectation is the best ceiling that Charlie can put on Ruth's expected winnings.

Chapter Terms

$E_C(q)$	46	Minimax expectation	49
Minimax strategy	49		

EXERCISES 5

For each of the games in Exercises 1–15,
   a) draw the graph of $E_C(q)$,
   b) determine a minimax strategy,
   c) find Ruth's minimax expectation,
   d*) specify for which values of q Ruth's optimal counterstrategy is $[1, 0]$ and for which it is $[0, 1]$.

1.

1	3
2	1

2.

3	2
0	2

3.

2	1
3	1

4.

2	1
1	3

5.

2	1
4	6

6.

1	−3
−2	1

7.

1	3
−2	1

8.

0	−1
0	1

9.

2	4
0	−1

10.

1	1
2	−1

11.

1	1
−3	1

12.

−2	1
0	3

13.

5	−1
−4	3

14.

−1	3
6	−4

15.

8	2
2	−8

SOLUTIONS OF ZERO-SUM GAMES

The solution of an $m \times n$ zero-sum game is defined and
a method for finding it in the 2×2 case is provided.

The examples of the previous two chapters contain an apparent coincidence. For each of those games Ruth's maximin expectation (computed in Chapter 4) and her minimax expectation (computed in Chapter 5) are equal. For example, in Bombing-sorties both expectations turn out to be $83.\overline{3}\%$. Table 1 below lists the common value v of both types of expectations for all of these examples. This is a unexpected coincidence because these two expectations, and their accompanying guarantees, have definitions that are diffrent in essential ways. The maximin expectation is a floor, or minimum, that Ruth guarantees for her expected winnings by using the maximin strategy, whereas the minimax expectation is a ceiling that Charlie can impose on Ruth's expectation by employing his minimax strategy. Since the maximin and minimax strategies are in general different—they are $[2/3, 1/3]$ and $[5/6, 1/6]$ in Bombing-sorties—there is no obvious reason why the maximin and minimax expectations should always be the same.

In order to further impress the reader with the surprising nature of this coincidence we pause to reexamine nonrepeated games. Guarantees can be made for these games too, but Ruth's and Charlie's guarantees will in general be very different. For example, in Bombing-sorties, if Ruth sends out only one mission, she can guarantee an 80% likelihood of success by placing the bomb on the bomber. Charlie, on the other hand, can guarantee by attacking that same bomber that Ruth's likelihood of success will not exceed 90%. Similarly, in any single play of the abstract game below, Ruth can guarantee a payoff of at least

| 1 | 8 |
| 4 | 3 |

3 by selecting the second row, while Charlie can hold Ruth's winnings down to no more than 4 by selecting the first column. Thus, for nonrepeated games, floor and ceiling type guarantees need not agree.

The coincidence of the guarantees for repeated zero sum games is the central theorem of Game Theory. It will be first stated and discussed in the context of 2×2 games and reformulated in a more general context later in the chapter.

THEOREM 1. *For every* 2×2 *zero-sum game there is a single number* v *such that*

i) *the maximin strategy guarantees Ruth an expected payoff of at least* v;
ii) *the minimax strategy guarantees Charlie that Ruth's expected payoff will not exceed* v.

The nature of the guarantees is such that if both players employ their recommended strategies, then the expected payoff will be exactly v. In other words, when Ruth employs the maximin strategy and Charlie employs the minimax strategy, then Ruth can expect to win v units per play (on the average). For this reason the number v is called the *value* of the game. The value of a game together with its maximin and minimax strategies constitute the *solution* of the game. Thus, by Example 1 of the previous two chapters the solution of Penny-matching is

$$\text{value} = 0$$
$$\text{maximin strategy} = [.5, .5]$$
$$\text{minimax strategy} = [.5, .5].$$

Similarly, Example 2 of the previous two chapters is summarized by saying that the solution of Bombing-sorties is

$$\text{value} = 83.\overline{3}\%$$
$$\text{maximin strategy} = [2/3, 1/3]$$
$$\text{minimax strategy} = [5/6, 1/6].$$

The previous two chapters detailed methods for finding the solution of any 2×2 zero-sum game. We now proceed to provide some shortcuts. For this purpose it is convenient to classify the 2×2 games into two types.

1) Strictly determined 2×2 **games**—those for which Ruth has a pure maximin strategy (and, as it turns out, Charlie has a pure minimax strategy). Graphically, this means that the line segments corresponding to $r_1(p)$ and $r_2(p)$ either do not intersect at all (Example 4.5), intersect in a common endpoint, or else their intersection point is not **the** highest point on the graph of $E_R(p)$ (Example 4.4). The reason for this nomenclature is that the strategies being pure, each player knows with complete certainty which option he will play next— the one indicated by the pure strategy.

TABLE 6.1. Expected values of some games

Example	v
4.1, 5.1	0
4.2, 5.1	$83.\bar{3}\%$
4.4, 5.4	2
4.5, 5.5	1

2) Nonstrictly determined 2×2 **games**—all the other games, for which, necessarily, the maximin strategies (and, as it turns out, also the minimax strategies) are never pure. Graphically this means that the line segments corresponding to $r_1(p)$ and $r_2(p)$ intersect internally and the point of intersection is higher than any other point on the graph of $E_R(p)$. In other words, these line segments form an approximate figure X. Such is the case for both Penny-matching and Bombing-sorties.

Strictly determined games have a convenient structural characterization that makes them easy to recognize.

THEOREM 2. *A* 2×2 *zero-sum game is strictly determined if and only if it contains an entry s which is minimal for its row and maximal for its column.*

An entry in a 2×2 game that is minimal for its row and maximal for its column is called a *saddle point*. Thus, the entry 2 is a saddle point of the game

0	-1
2	3

of Example 4 of Chapters 4, 5. Similarly, the entry 1 is a saddle point of the game

3	1
-1	-2

of Example 5 of Chapters 4, 5. Notice that these entries also constitute the values of these games, and that is no coincidence.

THEOREM 3. *The saddle point of a strictly determined 2×2 game is also its value, and its row and column constitute pure maximin and minimax strategies.*

The following procedure takes all the guesswork and/or graphing out of the task of recognizing and solving strictly determined games. Given a game, write at the bottom of each column that column's maximum entry and write at the right of each row that row's minimum entry (see Table 2). If any of the column maxima equals any of the row minima, the game is strictly determined, that common entry is the saddle point and the value of the game, and its row and

TABLE 6.2. Three games.

a				b				c		
5	3	$\underline{3}$		2	3	2		-2	0	-2
4	1	1		4	1	1		-1	1	$\underline{-1}$
5	$\underline{3}$			4	3			$\underline{-1}$	1	

value = 3	Not	value = -1
	strictly	
maximin	determined	maximin
strategy = $[1, 0]$		strategy = $[0, 1]$
minimax		minimax
strategy = $[0, 1]$		strategy = $[1, 0]$

column constitute the respective pure maximin and minimax strategies of the game. This is the case for games a and c of Table 2. Game b, however, is nonstrictly determined.

The 2×2 nonstrictly determined zero-sum games are subject to a solution procedure that is just as simple as the one that solves the strictly determined variety. For any such game

a	b
c	d

$$(1)$$

we define *Ruth's oddments* to be that one of the pairs

$$[d - c, a - b] \quad \text{or} \quad [c - d, b - a]$$

which consists of positive numbers alone (the game being nonstrictly determined, one of them will have this property). For example, for game b of Table 2 Ruth's oddments are

$$[4 - 1, 3 - 2] = [3, 1]$$

whereas for the (nonstrictly determined) game

5	-2
1	4

Ruth's oddments are

$$[4 - 1, 5 - (-2)] = [3, 7].$$

The maximin strategy is obtained from the oddments when they are each divided by their sum. In game b of Table 6.2, the maximin strategy is

$$\left[\frac{3}{3+1}, \frac{1}{3+1}\right] = \left[\frac{3}{4}, \frac{1}{4}\right] = [.75, .25]$$

and the oddments $[3, 7]$ of the game above yield the maximin strategy

$$\left[\frac{3}{3+7}, \frac{7}{3+7}\right] = \left[\frac{3}{10}, \frac{7}{10}\right] = [.3, .7].$$

Charlie's oddments in game (1) consist of that one of the pairs

$$[d - b, a - c] \quad \text{or} \quad [b - d, c - a]$$

which consists of two positive numbers. These oddments are converted to the minimax strategy in the same manner as above. Thus, for game b of Table 6.2 Charlie's oddments are

$$[3 - 1, 4 - 2] = [2, 2]$$

and his minimax strategy is

$$\left[\frac{2}{2+2}, \frac{2}{2+2}\right] = \left[\frac{2}{4}, \frac{2}{4}\right] = [.5, .5].$$

The rationale for this procedure is given in Lemmas 9, 10 below. The value of a **nonstrictly** determined 2×2 zero-sum game can be computed from the auxiliary diagram that is based on maximin and minimax strategies. For game b of Table 6.2 this diagram is

	.5	.5
.75	2	3
.25	4	1

and so the value of the game is

$$.75 \times .5 \times 2 + .75 \times .5 \times 3 + .25 \times .5 \times 4 + .25 \times .5 \times 1$$
$$= .75 + 1.125 + .5 + .125 = 2.5.$$

The following two examples should help in pulling the various methods of this section together.

EXAMPLE 4. Solve the game

0	−3
4	1

We first check to see whether this game is strictly determined, and that indeed

		0
0	−3	
4	1	$\underline{1}$

4    $\underline{1}$

turns out to be the case. Consequently, this game has value 1, Ruth's maximin strategy is $[0,1]$ and Charlie's minimax strategy is $[0,1]$.

EXAMPLE 5. Solve the game

0	3
4	1

The diagram below shows that this game is nonstrictly determined.

		0
0	3	
4	1	1

4   3

Consequently, Ruth's oddments are $[4-1, 3-0] = [3,3]$ and her maximin strategy is $[3/6, 3/6] = [.5, .5]$. Charlie's oddments are $[3-1, 4-0] = [2,4]$ and his minimax strategy is $[2/6, 4/6] = [1/3, 2/3]$. The value of the game comes from the auxiliary diagram

	1/3	2/3
.5	0	3
.5	4	1

and equals

$$.5 \times \frac{1}{3} \times 0 + .5 \times \frac{2}{3} \times 3 + .5 \times \frac{1}{3} \times 4 + .5 \times \frac{2}{3} \times 1$$
$$= 0 + 1 + \frac{2}{3} + \frac{1}{3} = 2.$$

The calculation of the value of a nonstrictly determined game can be further simplified by observing that since the peak point of the graph of $E_R(p)$ lies on both the graphs of $r_1(p)$ and $r_2(p)$ it follows that either of the pure strategies $[1,0]$ or $[0,1]$ can be substituted for Charlie's minimax strategy. In other words, the auxiliary diagram of Example 5 can be replaced by either of the diagrams

	1	0
.5	0	3
.5	4	1

	0	1
.5	0	3
.5	4	1

Indeed, these diagrams yield respectively the values

$$.5 \times 1 \times 0 + .5 \times 1 \times 4 = 0 + 2 = 2$$

and

$$.5 \times 1 \times 3 + .5 \times 1 \times 1 = 1.5 + .5 = 2,$$

in agreement with the previously derived value. Similarly, Ruth's maximin strategy could have been replaced with either $[1, 0]$ or $[0, 1]$, leading to the auxiliary diagrams

	1/3	2/3
1	0	3
0	4	1

	1/3	2/3
0	0	3
1	4	1

and again the values

$$1 \times \frac{1}{3} \times 0 + 1 \times \frac{2}{3} \times 3 = 0 + 2 = 2$$

and

$$1 \times \frac{1}{3} \times 4 + 1 \times \frac{2}{3} \times 1 = \frac{4}{3} + \frac{2}{3} = \frac{6}{3} = 2.$$

As promised above, we now give the full statement of von Neumann's Minimax Theorem.

MINIMAX THEOREM 6. *For every $m \times n$ zero-sum game there is a number v which has the following properties:*

a) *Ruth has a mixed strategy that guarantees her an expected payoff of **at least** v;*

b) *Charlie has a mixed strategy that guarantees that (Ruth's) expected payoff will be **at most** v.*

The quantity v whose existence is asserted in Theorem 6 is called the *value* of the game. The strategies mentioned in parts *a* and *b* of this theorem are called, respectively, the *maximin* and the *minimax* strategies of the game. The value of the game together with its maximin and minimax strategies constitute its *solution*. This terminology is of course consistent with the way these words were used above for the 2×2 games.

Efficient computerized methods for finding both the maximin and minimax strategies and the value of a game are known. These, however, fall outside the scope of this book. Instead, the remainder of this chapter and Chapters 7–9 are devoted to the solution of some special cases.

The notion of a strictly determined 2×2 game extends to larger dimensions with no difficulty. An entry in an $m \times n$ game that is minimal for its row and maximal for its column is called a *saddlepoint*. Thus the entries 2 and -1 are respective saddlepoints of the two games of Table 3. The same bookkeeping method of comparing row minima with column maxima will locate these saddlepoints in the larger games just as well as it did for the 2×2 games. A game

TABLE 6.3. Two strictly determined games.

a				b					
9	−2	−5	−5	5	−2	1	−3	4	−3
5	1	−9	−9	0	−2	5	0	−1	−2
3	2	5	2	9	−1	0	2	1	−1
−5	0	1	−5	9	−1	5	2	4	
9	2	5							

with a saddlepoint is said to be *strictly determined*. The saddlepoint's entry is the value of the game and its row and column constitute Ruth's maximin and Charlie's minimax strategies. Thus, the solutions of the two games of Table 3 are

Game	a	b
Value	2	−1
Maximin strategy	$[0,0,1,0]$	$[0,0,1]$
Minimax strategy	$[0,1,0]$	$[0,1,0,0,0,]$

Proofs*

The formal proof of the minimax theorem for 2×2 games is preceded by an incomplete visual argument that lends support to its validity. Let p and q be replaced by x and y respectively, and let the expected value of the game

$$G = \begin{array}{|c|c|} \hline a & b \\ \hline c & d \\ \hline \end{array}$$

when Ruth and Charlie employ the mixed strategies $[1-x,x]$ and $[1-y,y]$, respectively be denoted by $E(x,y)$. Then this expectation function has the value

$$E(x,y) = z = (a-b-c+d)xy + (c-a)x + (b-a)y + a.$$

This function has as its graph a surface over the unit square $0 \leq x,y \leq 1$. When Ruth selects a strategy $[1-p,p]$, this is tantamount to intersecting the surface of $E(x,y)$ with the vertical plane $x = p$. The resulting cross section happens to be a straight line with parametric equations

$$x = p, \quad y = t, \quad z = [(a-b-c+d)p + (b-a)]t + (c-a)p + a. \tag{1}$$

The significance of this cross section is that as long as Ruth sticks to her specific choice of $[1-p,p]$, Charlie's choice of any q yields Ruth an expectation that

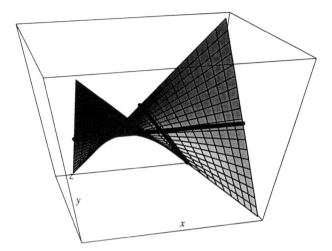

FIGURE 6.1. An expectation surface.

is equal to the z-coordinate of the corresponding point on the cross section. In particular, if this cross section happens to be horizontal, then Ruth's expectation will remain unaltered by any of Charlie's vacillations. Similarly, Charlie's selection of a strategy $[1 - q, q]$ is tantamount to intersecting the surface of $E(x, y)$ with the vertical plane $y = q$. This cross section too is a straight line with parametric equations

$$x = t, \quad y = q, \quad z = [(a - b - c + d)q + (c - a)]t + (b - a)q + a. \qquad (2)$$

This straight line has the same significance for Charlie as the previous one has for Ruth. For example, Figure 1 displays the expectation graph for the Penny Matching game. As is customary in all such displays, the grid on the surface consists of cross sections of the surface by planes parallel to the $x - z$ and $y - z$ planes. What is particular to this (and to all expectation surfaces) is that these grid lines are in fact straight lines. All of these cross sections have varying slopes, but those corresponding to $x = .5$ and $y = .5$, which have been designated by solid lines, are horizontal and so they must correspond to the minimax and maximin strategies. The fact that these two cross sections (necessarily) intersect corresponds to the equality of the maximin and minimax values of the players, and the z-coordinate of the intersection is this common value.

That the situation described in Figure 1 is fairly typical as long as $a - b - c + d \neq 0$ follows from $(1, 2)$ and the fact that if $\alpha \neq 0$ then any surface of the type

$$z = \alpha xy + \beta x + \gamma y + \delta$$

can be converted to a surface of the type

$$z = \alpha x'y' + \delta'$$

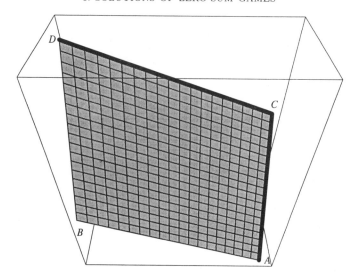

FIGURE 6.2. An expectation surface

by means of the straightforward translation of coordinates $x = x' - (\gamma/\alpha)$ and $y = y' - (\beta/\alpha)$. These last surfaces are easily seen to have a shape that is essentially the same as that of the surface in Figure 1.

In those cases where $a - b - c + d = 0$ the expectation surface has an equation of the type

$$z = \beta x + \gamma y + \delta$$

which is the portion of a plane that lies over the unit square as depicted in Figure 2. Once again the grid lines of the graph are straight lines whose significance is the same as before. None of these cross sections are horizontal, though, and so a different argument is needed to prove the Minimax Theorem. In the generic subcase we may suppose that the corners A, B, C, D in Figure 2 are each lower than the next. Suppose further that the cross sections that correspond to Ruth's fixing on a strategy $[1 - p, p]$ are those that are parallel to AB and CD (see Exercise 63). The maximin strategy then dictates that Ruth should choose that cross section whose lower end is as high as possible, i.e., CD. By fixing on a strategy Charlie chooses among the cross sections parallel to AC and BD. The minimax strategy dictates that he should choose a cross section whose higher end is as low as possible. That is AC. The coincidence of C on both of these choices is the purport of the Minimax Theorem and the z-coordinate of C is the value of the game. A similar argument disposes of all the other subcases (see Exercise 64). It is clear from the fact that the minimax value of the game (the height of C) occurs at the end of both the determining cross sections that in this case the game is strictly determined.

The formal proof of the Minimax Theorem for 2×2 games is broken down into a sequence of lemmas. In the geometrical proofs of these lemmas it is helpful, when drawing the graph of $E_R(p)$ (or $E_C(q)$), also to draw in the graph of the line $p = 1$ (or $q = 1$).

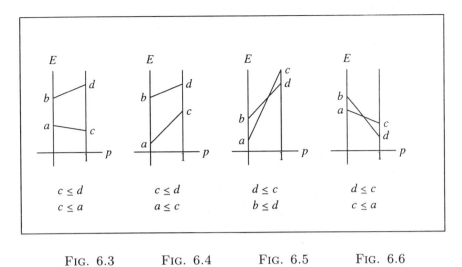

FIG. 6.3   FIG. 6.4   FIG. 6.5   FIG. 6.6

LEMMA 7. *Ruth has a pure maximin strategy if and only if G has a saddle point. In that case the payoff of the saddle point equals the maximin value.*

PROOF. It may be assumed without loss of generality that $a \leq b$. Suppose first that Ruth has a pure maximin strategy. The reader is reminded that the line segments $r_1(p) = (1-p)a + pc$ and $r_2(p) = (1-p)b + pd$, $0 \leq p \leq 1$, join the points $(0,a)$ to $(1,c)$ and $(0,b)$ to $(1,d)$ respectively. Since the maximin strategy is given by the high point of $E_R(p)$ it follows from the assumption of its purity that either these line segments do not intersect in an interior point (Figures 3, 4) or they both have nonnegative slopes (Figure 5) or they both have nonpositive slopes (Figure 6). Keeping in mind that $a \leq b$ it is now easily verified that in these four cases the game G has saddle points at a, c, d, a respectively and that these payoffs equal the maximin value of the game.

Conversely, suppose the game G has a saddle point. This saddle point entails inequalities on the payoffs of G which in turn yield partial information about the graphs of $r_1(p)$ and $r_2(p)$. These inequalities and their graphical implications are displayed in Figures 7–10. It is now easily verified that in each case Ruth has a pure maximin strategy with a value that coincides with the payoff of the saddle point.   q.e.d.

The next lemma should now come as no surprise and its proof is relegated to Exercise 65.

LEMMA 8. *Charlie has a pure minimax strategy if and only if G has a saddle point. In that case the payoff of the saddle point equals the minimax value.*

Between them, Lemmas 5 and 6 imply the validity of the Minimax Theorem for 2×2 games with saddle points.

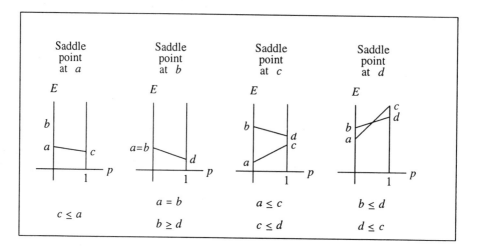

FIG. 6.7   FIG. 6.8   FIG. 6.9   FIG. 6.10

LEMMA 9. *If the game G has no pure maximin strategies then it has*

$$maximin\ strategy = \left[\frac{d-c}{a-b-c+d}, \frac{a-b}{a-b-c+d}\right]$$

$$maximin\ value = \frac{ad-bc}{a-b-c+d}.$$

PROOF. Since G is assumed to not have a pure maximin strategy, it follows that the graphs of $r_1(p)$ and $r_2(p)$ must intersect and have slopes of opposite signs. Thus the game has a maximin strategy that comes from the intersection of the graphs of $r_1(p)$ and $r_2(p)$. As $r_1(p) = (1-p)a+pc$ and $r_2(p) = (1-p)b+pd$ this nonpure maximin strategy is found by solving the equation

$$(1-p)a + pc = (1-p)b + pd$$

or

$$(a-b-c+d)p = a-b$$

for p. Since G is nonstrictly determined $a - b - c + d \neq 0$ (see Exercise 62), and so $p = (a-b)/(a-b-c+d)$ and $1-p = (d-c)/(a-b-c+d)$. The maximin value is then obtained by substituting this value of p into either $r_1(p)$ or $r_2(p)$. q.e.d.

Once again the next lemma comes as no surprise and its proof is relegated to Exercise 66.

LEMMA 10. *If the game G has no pure minimax strategies then it has*

$$minimax\ strategy = \left[\frac{d-b}{a-b-c+d}, \frac{a-c}{a-b-c+d}\right]$$

$$minimax\ value = \frac{ad-bc}{a-b-c+d}.$$

We are now ready to prove the Minimax Theorem for 2×2 games.

2×2 MINIMAX THEOREM 11. *For every 2×2 zero-sum game there is a number v which has the following properties:*

a) *Ruth has a mixed strategy that guarantees her an expected payoff of **at least** v;*

b) *Charlie has a mixed strategy that guarantees that Ruth's expected payoff will be **at most** v.*

PROOF. As noted above, if the game has a saddle point then this theorem follows from Lemmas 7 and 8. If the game does not have a saddle point then it follows from Lemmas 7 and 8 that the game has neither a pure maximin nor a pure minimax strategy. Hence, by Lemma 9 Ruth has a strategy that guarantees her an expected payoff of $(ad - bc)/(a - b - c + d)$ and by Lemma 10 Charlie has a strategy that guarantees that Ruth's expected payoff will not exceed $(ad - bc)/(a - b - c + d)$. Thus, $v = (ad - bc)/(a - b - c + d)$.   q.e.d.

Chapter Summary

Von Neumann's Minimax Theorem states that the guarantees provided by the maximin and minimax strategies for any $m \times n$ zero-sum game coincide numerically at a number called the value of the game. The value of the game together with its maximin and minimax strategies constitute the solution of the game. Quick methods are provided for solving 2×2 zero-sum games, but the solution of the general $m \times n$ game falls outside the scope of this book. Zero-sum games are classified into two varieties: strictly determined games whose maximin and minimax strategies are pure, and nonstrictly determined games.

Chapter Terms

Maximin strategy	54	Minimax strategy	54
Minimax Theorem	54, 59, 65	Nonstrictly determined game	55
Oddments	56	Saddle point	55, 59
Solution of game	54, 59	Strictly determined game	54
Value of game	54, 59		

EXERCISES 6

Solve the games in Exercises 1–40. I.e., find the value and both a maximin and a minimax strategy.

1.
1	3
2	1

2.
3	2
0	2

3.
2	1
3	1

4.
2	1
1	3

5.
2	1
4	6

6.
1	-3
-2	1

7.

1	3
-2	0

8.

0	-1
0	1

9.

2	4
0	-1

10.

1	1
2	-1

11.

1	1
-3	1

12.

-2	1
0	3

13.

5	-1
-4	3

14.

-1	3
6	-4

15.

8	2
2	-8

16.

3	1
1	3

17.

-3	1
-4	-2

18.

3	-3
-6	1

19.

0	1
0	-2

20.

-1	-2
1	3

21.

-1	-2
0	-3

22.

8	-2
2	8

23.

2	4
0	-1

24.

2	-2
2	2

25.

3	-1
0	3

26.

1	0
2	0

27.

1	-2
3	0

28.

2	1
2	4

29.

0	-1
-1	1

30.

-2	2
5	-5

31.

3	1	2
5	-1	3

32.

0	-2	7	2
1	-1	-1	2

33.

0	2
1	-1
2	3
-1	-2

34.

-5	5	0
2	-2	1
4	3	2

35.

1	-2	0	-3
-1	-2	0	1
5	-4	6	-3

36.

1	2	3
0	1	2
−1	0	1
−2	−1	0

37.

0	2	−4	0
−2	0	−2	4
4	2	0	6
0	0	−6	0

38.

−1	−2	0	1	0
3	4	2	2	4
2	0	1	−1	−3
1	−1	5	2	6

39.

2	1	1	3
6	−1	−7	8
−8	1	7	−6
5	0	−5	1
−1	−2	2	3

40.

5	2	−5	6	0
−2	0	2	0	−3
5	1	0	−1	−2
3	0	−1	2	−2
0	2	−1	0	−1

For which values of x are the games in Exercises 41–52 strictly determined?

41*.

1	x
0	2

42*.

2	x
0	1

43*.

1	x
0	1

44*.

x	2
1	0

45*.

x	2
1	x

46*.

x	x
1	0

47*.

x	x
x	0

48*.

x	x
x	x

49*.

1	2	x
7	0	3
6	5	4

1	2	0
7	x	3
6	5	4

50*.

1	2	6
7	0	3
x	5	4

51*.

1	2	x
7	x	3
x	5	4

52*.

53. Solve the game in Exercise 2.12.

54. Solve the game in Exercise 2.13.

55. Solve the game in Exercise 2.15.

56. Solve the game in Exercise 2.16.

57*. Prove that if a, b, c, d, x are any numbers then the two games below have the same maximin and the same minimax strategies. What is the relationship between their values?

a	b
c	d

$a + x$	$b + x$
$c + x$	$d + x$

58. (Dixit & Nalebuff) Ruth is a professional tennis player. She finds that when she anticipates her opponent's serve to be aimed at her forehand, she returns 90% of his serves when she is correct, but only 30% when she is wrong. On the other hand, when she anticipates the serve to be aimed at her backhand, she returns 60% of the serves when she is correct and only 30% when she is wrong. What is Ruth's maximin strategy? Suppose she is playing an opponent who is aware of these statistics; what is the opponent's minimax strategy?

59*. Is it true that the game

a	b
c	d

is strictly determined if and only if the game

a	c
b	d

is strictly determined? Prove your answer.

60*. Is it true that the game

a	b	c
d	e	f
g	h	i

is strictly determined if and only if the game

a	d	g
b	e	h
c	f	i

is strictly determined? Prove your answer.

61*. Is it true that the game

0	b	c
-b	0	d
-c	-d	0

is strictly determined if and only if the game

0	-b	-c
b	0	-d
c	d	0

is strictly determined? Prove your answer.

62*. Prove that if $a - b - c + d = 0$ then the game

a	b
c	d

is strictly determined.

63*. Modify the informal proof of the 2×2 Minimax Theorem to cover the case where the cross sections that correspond to fixing $x = p$ are parallel to AC and BD.

64*. Complete the informal proof of the 2×2 Minimax Theorem when the z coordinates of the corners A, B, C, D in Figure 2 are not all distinct.

65*. Prove Lemma 8.

66*. Prove Lemma 10.

A game is said to be *fair* if its value is 0.

67*. For which values of a, b, c is the game

a	0	0
0	b	0
0	0	c

fair? Justify your answer.

68*. For which values of a, b is the game

1	a
b	-1

fair? Justify your answer.

69*. For which values of a, b, c is the game

a	0
b	c

fair? Justify your answer.

70*. Prove that if an $m \times n$ game has more than one saddle points then their payoffs are all equal.

7

2 x n AND m x 2 GAMES

**The solution of zero-sum games in which
one of the players has only two options is given.**

The graphical method of Chapter 4 can be used to solve any $2 \times n$ game that is nonstrictly determined. It is only necessary to bear in mind that in such games Charlie has to select from n available pure strategies rather than just 2. Accordingly, if, for $j = 1, 2, \ldots, n$, $r_j(p)$ denotes the expected payoff when Ruth employs strategy $[1 - p, p]$, $0 \leq p \leq 1$, and Charlie consistently selects the j-th column, and if $E_R(p)$ again denotes the payoff Ruth can reasonably expect when employing strategy $[1 - p, p]$, then the analog of (1) of Chapter 4 is

$$E_R(p) = \text{the minimum of } \{r_1(p), r_2(p), \ldots, r_n(p)\}$$

and the graph of $E_R(p)$ coincides for each p with the lowest of the points on the graphs of $r_1(p), r_2(p), \ldots, r_n(p)$. As was the case before, if the j-th column of the game is

g
h

then the graph of $r_j(p)$ consists of the line segment that joins the points $(0, g)$ and $(1, h)$.

EXAMPLE 1. Solve the game

-2	0	-1	2
3	1	0	-1

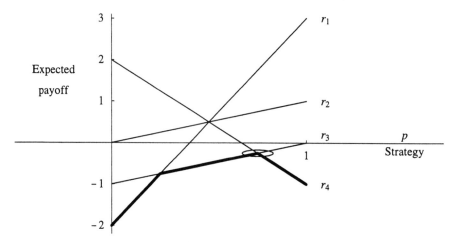

FIGURE 7.1. A graph of Ruth's expectation.

The row and column bookkeeping below testify that this game is nonstrictly

−2	0	−1	2	−2
3	1	0	−1	−1

3	1	0	2

determined. We may therefore pass on to the graphical method. The graph of Ruth's expected payoff is the broken heavy line of Figure 1. The peak of Ruth's graph, which determines Ruth's maximin strategy, is the intersection of the graphs of $r_3(p)$ and $r_4(p)$, and is circled in the figure. The corresponding value of p could of course be extracted from the equation $r_3(p) = r_4(p)$, but a modification of the *oddments* method works better here too. Note that the graphs of $r_1(p)$ and $r_2(p)$ both pass above the circled peak point. It therefore follows that their columns are irrelevant to Charlie's minimax strategy, since each of them would work to Ruth's clear benefit. Thus, Charlie's minimax strategy must have the form $[0, 0, 1 - q, q]$, for some $0 \le q \le 1$. Since columns 1 and 2 should never be used, they might as well be deleted from the original array, thus resulting in the reduced 2×2 subgame

−1	2
0	−1

This subgame does not have a saddlepoint either and so it may be solved by the oddments method. Here Ruth's oddments are $[1, 3]$ and so her maximin strategy for both the subgame and the original game is $[.25, .75]$. On the other hand, Charlie's oddments for the subgame are $[3, 1]$ and so his minimax strategy for the subgame is $[.75, .25]$ and his minimax strategy for the original game is

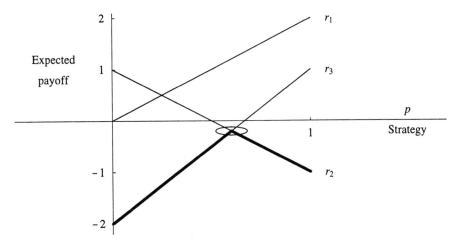

FIGURE 7.2. A graph of Ruth's expectation.

$[0, 0, .75, .25]$. The value of the original game is the same as the value of the subgame whose auxiliary diagram

	1	0
.25	−1	2
.75	0	−1

yields the answer

$$.25 \times 1 \times (-1) + .75 \times 1 \times 0 = -.25.$$

EXAMPLE 2. Solve the game

0	1	−2
2	−1	1

As this game is easily verified not to have a saddle point we go on to graph $E_R(p)$ in Figure 2. Inasmuch as the high point of this graph lies on the graphs of $r_2(p)$ and $r_3(p)$ it follows that we may restrict our attention to the subgame

1	−2
−1	1

that consists of the second and third columns of the original game. This subgame is nonstrictly determined and has oddments $[2, 3]$ for Ruth and $[3, 2]$ for Charlie. The original game has maximin strategy $[.4, .6]$, minimax strategy $[0, .6, .4]$, and the auxiliary diagram

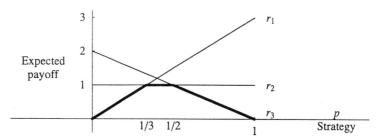

FIGURE 7.3. An unusual game.

	1	0
.4	1	−2
.6	−1	1

yields the value

$$v = .4 \times 1 \times 1 + .6 \times 1 \times (-1) = .4 - .6 = -.2$$

for the original game.

Surprisingly, it is possible for a **nonstrictly** determined $2 \times n$ game to reduce to a **strictly** determined subgame. This can happen only if the peak of the graph of $E_R(p)$ is actually a plateau and this exceptional case calls for caution.

EXAMPLE 3. The 2×3 game below clearly has no saddle point.

0	1	2	0
3	1	0	0
3	1	2	

For this game, the graph of $E_R(p)$ appears in Figure 3 and it has a plateau rather than a peak. If we were to choose to work (by analogy with the previous examples) with the intersection of the graphs of $r_1(p)$ and $r_2(p)$ we would obtain the subgame

0	1	0
3	1	1
3	1	

which clearly does possess a saddlepoint.

The given game has the additional interesting feature that its maximin strategy is not unique. Any point of the portion of the graph of $r_2(p)$ that lies between its intersections with the graphs of $r_1(p)$ and $r_3(p)$ yields a maximin strategy. These points of intersection can still be obtained by the oddments method, even though the subgames are strictly determined. Thus, the left endpoint corresponds to the subgame above which yields oddments $[3-1, 1-0] = [2, 1]$ and maximin strategy $[2/3, 1/3]$ for Ruth. The right endpoint corresponds to the subgame

1	2
1	0

in which Ruth's oddments are $[1-0, 2-1] = [1, 1]$ yielding a maximin strategy of $[.5, .5]$. Thus, any strategy $[1-p, p]$ with $1/3 \le p \le 1/2$ will serve as the maximin strategy. The value of the game is the value of $E_R(p)$ for these p, and a glance at Figure 3 tells us that this is the value 1 (the height of the plateau off the p-axis). Finally, the minimax strategy is that which guarantees to Charlie that Ruth's expected payoff will not exceed 1 and so it is the pure strategy that selects the (second) column that created the plateau, i.e., $[0, 1, 0]$.

The following theorem should be kept in mind when solving the exercises.

THEOREM 4. *When solving a $2 \times n$ zero-sum game, if a maximin strategy is determined by the point of intersection of two of the r_j's, then the corresponding value of p can be determined by the oddments method.*

Games of dimensions $m \times 2$ are subject to a resolution that is similar to that of $2 \times n$ games, the main difference being that now it is Charlie's point of view that guides us. Accordingly, if for each $i = 1, 2, \dots, m$ $c_i(q)$ denotes the expected payoff when Ruth consistently employs her i-th row against Charlie's arbitrary mixed strategy of $[1-q, q]$, $0 \le q \le 1$, and if $E_C(q)$ denotes the expected payoff that Ruth will select under these circumstances, then the analog of (1) of Chapter 5 is

$$E_C(q) = \text{the maximum of } \{c_1(q), c_2(q), \dots, c_m(q)\}.$$

The graph of $E_C(q)$ coincides for each q with the highest of the points on the graphs of $c_1(q), c_2(q), \dots, c_m(q)$. As was the case before, if the i-th row of the game is

g	h

then the graph of $c_i(q)$ consists of the line segment that joins the points $(0, g)$ and $(1, h)$.

EXAMPLE 5. The game

3	0
2	3
-1	4

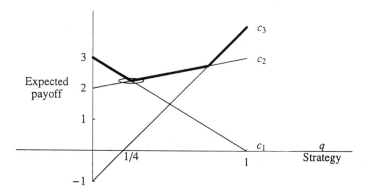

FIGURE 7.4. A graph of Charlie's expectation.

is nonstrictly determined and the graph of Charlie's expected payoff is the broken heavy line of Figure 4. In this figure the graph of $c_i(q)$ has been labeled on its right with c_i for each $i = 1, 2, 3$. The low point of Charlie's graph, which determines his minimax strategy and is circled in the figure, is the intersection of the graphs of $c_1(q)$ and $c_2(q)$. Consequently, it is possible to restrict our attention to the subgame

3	0
2	3

which consists of the first two rows of the given game. For this subgame Charlie's oddments are $[3 - 0, 3 - 2] = [3, 1]$ which yield $[.75, .25]$ as his minimax strategy. Ruth's oddments for the subgame are $[3 - 2, 3 - 0] = [1, 3]$ and her maximin strategy is therefore $[.25, .75, 0]$. The value of the given game is the same as the value of the subgame which is

$$1 \times .75 \times 3 + 1 \times .25 \times 0 = 2.25.$$

It is possible for a nonstrictly determined game to reduce to a strictly determined game, as is the case when the graph of $E_C(q)$ has a flat floor rather than a single low point. As indicated by Theorem 6 below, when this happens the endpoints of the floor can still be determined by the oddments method.

THEOREM 6. *When solving an $m \times 2$ zero-sum game, if a minimax strategy is determined by the point of intersection of two of the c_i's, then the corresponding value of q can be determined by the oddments method.*

EXAMPLE 7. Solve the game

1	1
−2	2
2	−2

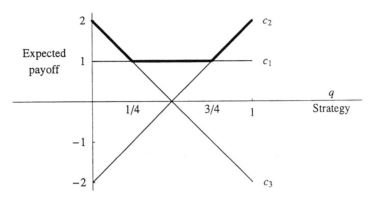

FIGURE 7.5. A graph of Charlie's expectation.

The graph of $E_C(q)$ is the broken heavy line in Figure 5.

The left endpoint of the flat bottom comes from Charlie's oddments in the subgame determined by the first and third rows:

1	1
2	−2

These oddments are $[1 - (-2), 2 - 1] = [3, 1]$ which yields a minimax strategy $[3/4, 1/4]$, or $q = 1/4$.

The right endpoint of the flat bottom comes from Charlie's oddments in the subgame determined by the first and second rows:

1	1
−2	2

These oddments are $[2 - 1, 1 - (-2)] = [1, 3]$ which yields a minimax strategy $[1/4, 3/4]$, or $q = 3/4$.

The value of the game is the value of $E_C(q)$ for any q between $1/4$ and $3/4$ and a glance at Figure 5 tells us that this is 1 (the height of the flat bottom of Charlie's expectation curve off the q-axis). Finally, the maximin strategy is any strategy of Ruth's that holds Charlie's expectation down to 1. This Ruth can accomplish by consistently choosing the first row, i.e., with the pure maximin strategy of $[1, 0, 0]$.

THE JAMAICAN FISHING VILLAGE. *This chapter's last example describes a game that was extracted by the anthropologist W. C. A. Davenport from his observations of the behavior of the inhabitants of a certain Jamaican fishing village. These fishermen possessed twenty six fishing canoes manned each by a captain and two or three crewmen. The fishing took the form of setting pots (traps) and drawing from them. The fishing grounds were divided into inside and outside banks. The inside banks lay from 5 to 15 miles offshore, whereas*

the outside banks lay beyond. The crucial factor that distinguished between the two areas was the occasional presence of very strong currents in the latter, which rendered fishing impossible. Accordingly, each captain must decide on a trap setting policy. He could:
(1) *set all his pots inside,*
(2) *set all the pots outside, or*
(3) *set some of the pots inside and some outside.*

Davenport modeled this situation as a 3×2 game whose players are the village and the environment. Each of the captains' selections of a fishing strategy constituted a play on the part of the village. The environment "decided" on its option by either sending a current or not. Based on his observations of the local market place and the costs accrued by the captains, Davenport estimated the payoffs of the various possibilities as follows:

| | | **Environment** | |
		Current	No-current
	Inside	17.3	11.5
Village	In-out	5.2	17.0
	Outside	−4.4	20.6

The monetary unit was the Pound and the reason for the one negative payoff was that the Captain must pay his crew for each outing, regardless of the catch. Davenport went on to treat this as a zero-sum two person game, solved it as such, and compared the game's maximin strategy to the actual distribution of the captains' choices.

Pretending that the environment is a conscious player, its expectation graph $E_C(q)$ is drawn in Figure 6. This figure indicates that the requisite subgame consists of the first two rows of the given game

17.3	11.5
5.2	17.0

The village's oddments are $[17.0 - 5.2, 17.3 - 11.5] = [11.8, 5.8]$ which, in turn, yield the maximin strategy of

$$[.67, .33, 0]$$

with a corresponding maximin expectation of

$$.67 \times 17.3 + .33 \times 5.2 = 13.31.$$

This is to be understood as saying that Game Theory's recommendation to the village is that 67% of its fishing should be done in the inside banks exclusively, 33% as an inside-outside combination, and none of the fishermen should dedicate themselves to fishing in the outside banks alone. This way the village can guarantee its fishermen an expected payoff of at least 13.31 per outing.

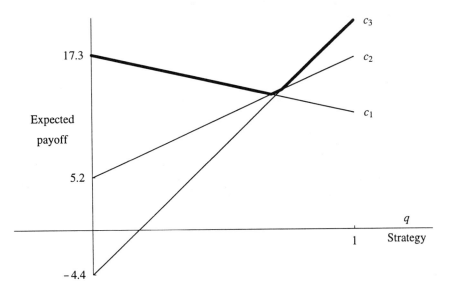

FIGURE 7.6. A fishing game.

Davenport observed that 18 (69%) of the captains fished only in the inside banks, 8 (31%) adopted the inside-outside combination, and none restricted their fishing to the outside banks alone. A remarkable fit between theory and observation.

Two additional comments may be in order here. During his two year stay Davenport observed that a current was present in the outside bank on 25% of the days. This can be interpreted as a fixed strategy of $[.25, .75]$ for the environment. In accordance with Chapter 3, the village's optimal counterstrategy should be a pure strategy whose specific value is computed on the basis of the auxiliary diagrams

	.25	.75
1	17.3	11.5
0	5.2	17.0
0	−4.4	20.6

a

	.25	.75
0	17.3	11.5
1	5.2	17.0
0	−4.4	20.6

b

	.25	.75
0	17.3	11.5
0	5.2	17.0
1	−4.4	20.6

c

which have respective expected payoffs v_a, v_b, v_c, where

$$v_a = 1 \times .25 \times 17.3 + 1 \times .75 \times 11.5 = 12.95$$
$$v_b = 1 \times .25 \times 5.2 + 1 \times .75 \times 17.0 = 14.05$$
$$v_c = 1 \times .25 \times (-4.4) + 1 \times .75 \times 20.6 = 14.35.$$

Since 14.35 is the largest of these expected payoffs, the village as a whole would profit in the long run by restricting their fishing to the outside banks. Nevertheless, such risk taking seems to be contrary to human nature.

Something needs to be said about the minimax strategy of this game. The environment's oddments are $[17.0 - 11.5, 17.3 - 5.2] = [5.5, 12.1]$ and so its minimax strategy is $[.31, .69]$. In other words, if the environment were a sentient being bent on yielding as little as possible to the Jamaican fishermen, it would be inclined to create a current on 31% of the days. This is not as exciting a fit with the observed 25% as we had for the village's maximin strategy, but it is still close. While this is not to be taken as evidence for the anthropomorphic view of nature, it should be pointed out that the game's payoffs are dependent on many factors, including the frequencies of the currents and the village's fishing strategies. Is it possible that this dependence might cause the game's entries to stabilize at values whose minimax strategy agrees with the actual frequency of the current?

Chapter Summary

The graphical method of Chapters 4, 5 was applied to the solution of all zero-sum games in which one of the players has exactly two options. As part of this process it became evident that every such game contains a 2×2 subgame whose solution gives the value and maximin and minimax strategies of the original game. This technique was then applied to the game theoretic analysis of the fishing strategies adopted by the members of a Jamaican fishing village.

EXERCISES 7

Solve the games in Exercises 1–23. I.e., find their values and some maximin and minimax strategies.

1.

−1	1	2
2	1	−2

2.

4	6	−3	5
5	−2	8	4

3.

4	−2	1
−2	5	−1

4.

1	5	3	4
5	1	3	2

5.

0	−2	5
0	6	−4

6.

2	1	3
5	0	4

7.

3	−7	0	−4
−6	1	−1	−5

8.

−1	−5	4	0
−1	5	−1	0

9.

2	3	2	4
1	5	−3	−4

10.

1	7	4	2
6	0	4	2

11.

-1	2
1	1
2	-2

12.

4	-2
-2	5
1	-1

13.

0	0
2	6
1	-4

14.

4	5
6	-2
-3	8
5	4

15.

1	5
2	6
3	4
3	0

16.

3	-6
-7	1
0	-1
4	-5

17.

2	1
3	5
2	-3
4	-4

18.

1	6
7	0
4	4
2	2

19.

-1	-1
-5	5
4	-1
0	0

20.

2	5
1	0
3	4

21.

1	5
5	1
3	3
4	2

22.

0	0
-2	6
5	-4

23.

1	1
1	1
1	1

24. Solve the game in Exercise 2.14.

DOMINANCE

**Some $m \times n$ games can be solved by identifying and
deleting irrelevant rows and columns.**

As was mentioned above, the general solution of $m \times n$ games lies outside the
scope of this text. Occasionally, however, larger games may be reduced to more
tractable dimensions by the deletion of some rows and/or columns whose irrel-
evance is easily recognized. This is first demonstrated with an example which
will be followed by a statement of the general principle in question.

EXAMPLE 1. Consider the game

−5	4	6
3	−2	2
2	−3	1

Observe that Ruth should never choose the third row since she could always do
better by choosing the second row, **no matter what Charlie's selection is**.
The reason for this is that every entry in the second row is greater than the entry
directly below it, and Ruth is always looking to get the larger payoff. Similarly,
Charlie should never choose the third column since each of its entries is larger
than the corresponding entry in the second column. This is dictated by Charlie's
goal of keeping the payoff (Ruth's gain) as small as possible. Consequently, each
of the players will restrict their attention to their first two options, thereby
reducing the given game to the subgame

-5	4
3	-2

which is easily solved by the oddments method.

The underlying idea here is that of dominance. Given two lists of equal length

$$L_1 = (a_1, a_2, \ldots, a_n) \quad \text{and} \quad L_2 = (b_1, b_2, \ldots, b_n)$$

we say that list L_1 *dominates* L_2 provided

$$a_i \geq b_i \quad \text{for all } i = 1, 2, \ldots, n.$$

Thus, the second row of the game of Example 1 dominates its third row because

$$3 \geq 2, \quad -2 \geq -3, \quad 2 \geq 1.$$

Similarly the second column of this game is dominated by its third column.

The following observation is justified by the fact that Ruth always looks for the larger payoff whereas Charlie seeks to minimize Ruth's gains.

THEOREM 2. *In any zero-sum game Ruth has a maximin strategy that does not employ any dominated rows, and Charlie has a minimax strategy that does not employ any dominating columns.*

EXAMPLE 3. Solve the 4×4 zero-sum game

3	-2	2	-1
1	-2	2	0
0	6	0	7
-1	5	0	8

Since this game is not strictly determined, we go on to look for dominance amongst its rows and columns. The fourth column dominates the second one, and this is in fact the only instance of dominance. This is denoted by

3	-2	2	-1
1	-2	2	0
0	6	0	7
-1	5	0	8

and the fourth column is deleted. The resulting 4×3 subgame has dominance amongst its rows:

3	- 2	2
1	- 2	2
0	6	0
- 1	5	0

After the dominated rows are deleted from this subgame the resulting 2×3 subgame still has some dominance

3	- 2	2
0	6	0

which finally reduces the original game to

−2	2
6	0

Ruth's oddments in this subgame are $[6 - 0, 2 - (-2)] = [6, 4]$. Since the rows of this 2×2 game are the remnants of the first and third rows of the original game, this latter game has the maximin strategy of $[.6, 0, .4, 0]$. Similarly, Charlie's oddments of $[2 - 0, 6 - (-2)] = [2, 8]$ yield the minimax strategy $[0, .2, .8, 0]$ for the original game. The value of the original game equals the value of its subgame. Using the auxiliary diagram

	0	1
.6	−2	2
.4	6	0

this common value is computed as $.6 \times 2 + .4 \times 0 = 1.2$.

The following example is due to J. D. Williams.

EXAMPLE 4. During a period of political uncertainty, an investor is weighing his options of buying government bonds, armament stocks, or industrial stocks. He sees the near future as bringing either actual war, cold war, or peace, and, on the basis of past experience, computes the following rates of interest for each eventuality (the percent signs are implicit):

| | **Future** | | |
	Actual war	Cold war	Peace
Government bonds	2	3	3.2
Armament stocks	18	6	−2
Industrial stocks	2	7	12

Investor (label for the three row categories)

Viewing this as a game between the investor and nature we now calculate the investor's maximin strategy.

Since the third row dominates the first one, the latter can be discarded. For the resulting 2×3 game

18	6	−2
2	7	12

the graph of the investor's (i.e., Ruth's) strategy is drawn in Figure 1. This graph yields the subgame

18	6
2	7

The investor's oddments for the final subgame are $[7-2, 18-6] = [5, 12]$ giving $[0, 5/17, 12/17]$ as the maximin strategy. In other words, the game theoretic point of view recommends that the investor ignore the government bonds, invest $5/17 = 29.4\%$ of his funds in armament stocks and the remaining 70.6% in industrial stocks. The value of the game is computed from the auxiliary diagram

	1	0
5/17	18	6
12/17	2	7

as

$$\frac{5}{17} \times 1 \times 18 + \frac{12}{17} \times 1 \times 2 = \frac{90 + 24}{17} = \frac{114}{17} = 6.7.$$

Thus, the above maximin strategy guarantees the investor an expected return of 6.7% on his investments.

We conclude with a comment about the relationship between dominance and saddlepoints. Briefly put, the reduction process described in this chapter cannot uncover new saddlepoints (see Exercise 12). In other words, if a game is not strictly determined but does have some dominance amongst its rows and columns, then the subgame resulting from the appropriate eliminations is also not strictly determined. Thus, a game needs to be scrutinized for a saddlepoint only once.

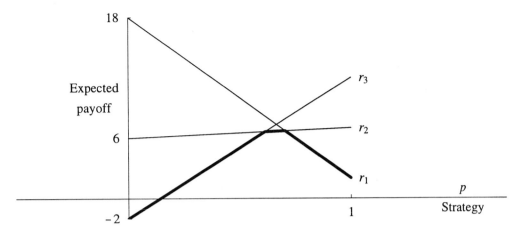

FIGURE 8.1. An investment game.

Chapter Summary

The concept of dominance can be used to identify options that are clearly "bad" for the player. Their deletion from an $m \times n$ game reduces its size and sometimes leads to a complete solution.

Chapter Terms

Dominance   84

EXERCISES 8

Solve the games in Exercises 1–10.

1.

3	3	4	1	2
3	3	3	2	2
0	-2	1	4	5

2.

3	2	0	2	-2
2	2	2	1	2
5	4	3	2	1

3.

2	3	2	4	5
-5	9	1	8	-1
8	-2	2	-5	3

4.

4	−2	3
6	8	−5
3	−3	3
5	8	−6
5	−2	4

5.

3	−3	−4
4	−2	−3
1	−1	−3
2	0	−2
5	−1	3

6.

5	−3	8
8	−2	5
6	−1	−3
2	0	2
−3	−1	6

7.

3	1	2	2	1
1	2	1	2	2
0	1	1	0	3

8.

2	0	1	1	0
0	1	0	1	1
−1	0	0	−1	2
−1	1	−1	1	0

9.

1	2	3	3
2	1	2	1
1	2	2	4
1	1	3	1
2	1	1	2

10.

1	1	2	2	2
2	3	3	2	1
1	2	1	2	2
2	2	4	2	1
1	3	1	2	1

11*. Prove that a 2 × 2 zero-sum game has a saddle point if and only if either one of its rows dominates the other, or else one of its columns dominates the other.

12*. Let **G** be a zero-sum game in which one row (column) dominates another, and let **H** be the game obtained from **G** by deleting the dominated (dominating) row (column). Prove that **G** is strictly determined if and only if **H** is strictly determined.

9

SYMMETRIC GAMES

Many games have a built in symmetry that makes the recognition of their solutions an easy matter.

Many concrete zero-sum games are symmetric in the sense that the players' roles are interchangeable. Such is the case for the Rock-scissors-paper and Morra games described in the first chapter. On the other hand, the Bombing-sorties and Jamaican fishing village games are not symmetric since the players have clearly distinct roles. The property of symmetry can sometimes be used to solve a game when the previous methods are of no avail.

Since in a symmetric game the players' roles are interchangeable, neither player has an advantage over the other and consequently neither player has any reason to expect to win in the long run. Moreover, any strategy that guarantees to Ruth a break-even expectation can also be used by Charlie for the same purpose. Hence we have the following theorem (see comments at the end of this chapter).

THEOREM 1. *Every symmetric zero-sum game has value 0 and identical maximin and minimax strategies.*

Games with value 0 are said to be *fair*. Unfortunately, as was seen in the game of Morra, knowing that a game is fair does not tell us how to find the maximin and minimax strategies. Nevertheless, this knowledge can be used to check whether any specific strategy, arrived at by either a random choice or a lucky guess, is or is not a maximin strategy. To wit, suppose $\mathbf{R} = [p_1, p_2, \ldots, p_m]$ is some strategy of Ruth's in some symmetric game $\mathbf{G}$. Suppose that Charlie possesses a pure strategy $\mathbf{C}$ which, when employed against $\mathbf{R}$ yields a negative expected payoff. Then it is clear that $\mathbf{R}$ is *not* a maximin strategy since, by definition, the maximin strategy *guarantees* an expected payoff of at least 0 against *any* strategy of Charlie's. On the other hand, if every **pure** response of Charlie's yields a nonnegative expected payoff, then the outcome of every **mixed** strategy of Charlie's, being a weighted average of the pure payoffs (see second proof of Theorem 3.5), is also nonnegative. Since 0 is the most that Ruth can reasonably expect here, it follows that this strategy **R** is in fact a maximin strategy for Ruth. This reasoning is summarized below.

THEOREM 2. *In a fair game, in order to verify that a given strategy of Ruth's is indeed maximin, it suffices to show that it guarantees a nonnegative expected payoff against each of Charlie's pure responses.*

EXAMPLE 3. We shall verify that the strategy $\mathbf{R} = [0, 4/7, 3/7, 0]$ is a maximin (and therefore also a minimax) strategy in the game of Morra. From the auxiliary diagrams

	1	0	0	0
0	0	2	−3	0
4/7	−2	0	0	3
3/7	3	0	0	−4
0	0	−3	4	0

a

	0	1	0	0
0	0	2	−3	0
4/7	−2	0	0	3
3/7	3	0	0	−4
0	0	−3	4	0

b

	0	0	1	0
0	0	2	−3	0
4/7	−2	0	0	3
3/7	3	0	0	−4
0	0	−3	4	0

c

	0	0	0	1
0	0	2	−3	0
4/7	−2	0	0	3
3/7	3	0	0	−4
0	0	−3	4	0

d

we compute that

$$v_a = \frac{4}{7} \times 1 \times (-2) + \frac{3}{7} \times 1 \times 3 = \frac{-8+9}{7} = \frac{1}{7},$$

$$v_b = \frac{4}{7} \times 1 \times 0 + \frac{3}{7} \times 1 \times 0 = 0 + 0 = 0,$$

$$v_c = \frac{4}{7} \times 1 \times 0 + \frac{3}{7} \times 1 \times 0 = 0 + 0 = 0,$$

$$v_d = \frac{4}{7} \times 1 \times 3 + \frac{3}{7} \times 1 \times (-4) = \frac{12-12}{7} = 0.$$

Since all these expected payoffs are nonnegative, it follows that the given strategy is indeed maximin.

The arrays associated with the symmetric games of Rock-scissors-paper and Morra possess a visual feature that is common to all symmetric games. All the entries on the diagonal from top left to bottom right are 0 and if a and b are entries in positions that are in mirror image location relative to this same diagonal, then $b = -a$. This follows from the interchangeability of the roles of the players and the fact that Ruth's gain is Charlie's loss (and vice versa). These properties are summarized as follows.

THEOREM 4. *The arrays of symmetric games are characterized by the property*

$$p_{i,j} = -p_{j,i} \quad i = 1, 2, \ldots, m, \quad j = 1, 2, \ldots, n,$$

where $p_{i,j}$ denotes the payoff in the i-th row and j-th column.

It follows that every 3×3 symmetric game has the form

0	a	b
$-a$	0	c
$-b$	$-c$	0

where a, b, c, are some real numbers. It so happens that these 3×3 symmetric games have solutions that are easily described. (See Exercise 12.)

THEOREM 5. *If the 3×3 symmetric game*

0	a	b
$-a$	0	c
$-b$	$-c$	0

is nonstrictly determined, then its maximin and minimax strategies are

$$\left[\frac{|c|}{|a| + |b| + |c|}, \quad \frac{|b|}{|a| + |b| + |c|}, \quad \frac{|a|}{|a| + |b| + |c|} \right].$$

EXAMPLE 6. For Rock-scissors-paper the array is

0	1	-1
-1	0	1
1	-1	0

and hence the maximin and minimax strategies are

$$\left[\frac{1}{1+1+1}, \quad \frac{1}{1+1+1}, \quad \frac{1}{1+1+1} \right] = \left[\frac{1}{3}, \quad \frac{1}{3}, \quad \frac{1}{3} \right],$$

just as expected.

EXAMPLE 7. The symmetric game

0	−2	−6
2	0	−4
6	4	0

has a saddle point in its lower right hand entry. Consequently its value is 0 and both its maximin and minimax strategies are $[0, 0, 1]$.

EXAMPLE 8. The symmetric game

0	−2	6
2	0	−4
−6	4	0

is nonstrictly determined. It has value 0 and both its maximin and minimax strategies are

$$\left[\frac{4}{2+6+4}, \frac{6}{2+6+4}, \frac{2}{2+6+4}\right] = \left[\frac{4}{12}, \frac{6}{12}, \frac{2}{12}\right] = \left[\frac{1}{3}, \frac{1}{2}, \frac{1}{6}\right].$$

EXAMPLE 9*. The symmetry of the game of Two-finger Morra can be used to derive all its maximin strategies. If $\mathbf{R} = [x, y, z, w]$ and $\mathbf{C} = [s, t, u, v]$ denote arbitrary strategies for Ruth and Charlie, then the auxiliary diagram

	s	t	u	v
x	0	2	−3	0
y	−2	0	0	3
z	3	0	0	−4
w	0	−3	4	0

yields the expected payoff

$$E = (-2y + 3z)s + (2x - 3w)t + (-3x + 4w)u + (3y - 4z)v.$$

If $\mathbf{R}$ is a maximin strategy, then the symmetry of the game implies that regardless of the values of s, t, u, v (as long as they are kept nonnegative), the value of E is nonnegative. This is logically equivalent to the following four inequalities:

$$-2y + 3z \geq 0, \quad 2x - 3w \geq 0, \quad -3x + 4w \geq 0, \quad 3y - 4z \geq 0$$

or

$$3z \geq 2y \quad \text{and} \quad 2x \geq 3w$$
$$3y \geq 4z \qquad\qquad 4w \geq 3x \qquad\qquad (1)$$

The two inequalities on the right can be combined into

$$8x \geq 12w \geq 9x,$$

and since x and w are nonnegative it follows that $x = w = 0$. The constraint $x + y + z + w = 1$ yields $z = 1 - y$. Substitution into the left hand inequalities of (1) then results in

$$3(1 - y) \geq 2y$$
$$3y \geq 4(1 - y)$$

or

$$\frac{3}{5} \geq y \geq \frac{4}{7} \qquad\qquad (2)$$

Hence any strategy of the form $[0, y, 1 - y, 0]$ is a maximin (and minimax) strategy for this game, as long as y is constrained by (2). For example, $[0, .6, .4, 0]$ and $\left[0, \frac{4}{7}, \frac{3}{7}, 0\right]$ are two such strategies.

We conclude with some remarks about Theorem 1. It has already been noted that the idea of modeling a game by means of a rectangular array goes back to Borel's papers of the early 1920's. The same papers also took the first steps towards the identification of the notion of a strategy with a sequence of probabilities and even a formulation of the notion of the maximin value of a game. This was all done in reference to symmetric games only. Such a restricted context is not surprising in view of the fact that the first games that come to mind usually possess a fair amount of symmetry.

It was taken for granted both at the beginning of this chapter and in the initial discussion of Penny-matching and Two-finger Morra that such symmetric games have strategies that guarantee the players an expected payoff of at least 0. This was not at all obvious to Borel at the time he initiated this study. In fact, at least for a while, he believed in the existence of symmetric games in which knowledge of the opponents strategy, no matter which one it might be, could always be turned to one's advantage. In other words, he did not believe in the existence of minimax and maximin strategies for these games. A surprising mistake that was shortly set straight by von Neumann's proof of the Minimax Theorem in 1928.

Chapter Summary

While the solution of symmetric zero-sum games is no easier than the solution of the general $m \times n$ zero-sum games, it is a simple matter to verify whether a proposed strategy for a given symmetric game is in fact a maximin (or minimax) strategy or not. A formula is given for the solution of any symmetric nonstrictly determined 3×3 zero-sum game.

Chapter Terms

EXERCISES 9

In each of Exercises 1–5 decide which of the given strategies $\mathbf{R}_1$, $\mathbf{R}_2$, $\mathbf{R}_3$, if any, is a maximin strategy for the given symmetric game $\mathbf{G}$.

1. $\mathbf{R}_1 = [4/7, 0, 1/7, 2/7]$, $\mathbf{R}_2 = [2/7, 4/7, 0, 1/7]$, $\mathbf{R}_3 = [0, 1/7, 2/7, 4/7]$,

$$G = \begin{array}{|c|c|c|c|}
\hline
0 & 2 & -2 & 1 \\
\hline
-2 & 0 & 1 & 2 \\
\hline
2 & -1 & 0 & -4 \\
\hline
-1 & -2 & 4 & 0 \\
\hline
\end{array}$$

2. $\mathbf{R}_1 = [3/8, 3/8, 0, 1/4]$, $\mathbf{R}_2 = [1/4, 3/8, 3/8, 0]$, $\mathbf{R}_3 = [3/8, 0, 3/8, 1/4]$,

$$G = \begin{array}{|c|c|c|c|}
\hline
0 & 1 & 2 & -3 \\
\hline
-1 & 0 & -1 & 2 \\
\hline
-2 & 1 & 0 & 3 \\
\hline
3 & -2 & -3 & 0 \\
\hline
\end{array}$$

3. $\mathbf{R}_1 = [1/4, 1/4, 1/4, 1/4]$, $\mathbf{R}_2 = [3/7, 1/7, 0, 3/7]$, $\mathbf{R}_3 = [4/7, 0, 0, 3/7]$,

$$G = \begin{array}{|c|c|c|c|}
\hline
0 & 3 & -2 & -1 \\
\hline
-3 & 0 & 0 & 3 \\
\hline
2 & 0 & 0 & -4 \\
\hline
1 & -3 & 4 & 0 \\
\hline
\end{array}$$

4. $\mathbf{R}_1 = [1/3, 0, 1/3, 1/3]$, $\mathbf{R}_2 = [1/3, 1/3, 1/3, 0]$, $\mathbf{R}_3 = [0, 1/3, 1/3, 1/3]$,

$$G = \begin{array}{|c|c|c|c|}
\hline
0 & 1 & -1 & -1 \\
\hline
-1 & 0 & 1 & -1 \\
\hline
1 & -1 & 0 & 1 \\
\hline
1 & 1 & -1 & 0 \\
\hline
\end{array}$$

5. $\mathbf{R}_1 = [.25, .5, .25, 0]$, $\mathbf{R}_2 = [.5, .25, .25, 0]$, $\mathbf{R}_3 = [.25, .25, .25, .25]$,

$$G = \begin{array}{|c|c|c|c|}
\hline
0 & 1 & -2 & 3 \\
\hline
-1 & 0 & 1 & -2 \\
\hline
2 & -1 & 0 & 1 \\
\hline
-3 & 2 & -1 & 0 \\
\hline
\end{array}$$

Solve the games in Exercises 6 – 10.

6.

0	1	−2
−1	0	−3
2	3	0

7.

0	1	2
−1	0	−3
−2	3	0

8.

0	−1	−2
1	0	3
2	−3	0

9.

0	−2	2
2	0	−3
−2	3	0

10.

0	3	−2
−3	0	2
2	−2	0

11*. Prove that the symmetric 3×3 game

0	a	b
$-a$	0	c
$-b$	$-c$	0

is strictly determined if and only if one of the following three conditions holds:
  i) $a \geq 0$ and $b \geq 0$,
  ii) $a \leq 0$ and $c \geq 0$,
  iii) $b \leq 0$ and $c \leq 0$.

12*. Prove Theorem 5.

13*. Let $a, b, c > 0$. The game below is an obvious generalization of the game of Morra.

0	a	$-b$	0
$-a$	0	0	b
b	0	0	$-c$
0	$-b$	c	0

  i) Find this game's maximin strategy when $a = 3$, $b = 4$, $c = 5$;
  ii) Find this game's maximin strategy when $a = 4$, $b = 3$, $c = 5$;
  iii) Find this game's maximin strategy when $a = 2$, $b = 4$, $c = 8$.

14*. Let $a, b, c > 0$. The game below is an obvious generalization of the game of Morra.

0	a	$-b$	0
$-a$	0	0	b
b	0	0	$-c$
0	$-b$	c	0

i) Show that if $b^2 = ac$, $r = \frac{b}{a} = \frac{c}{b}$, and α and β are two nonnegative numbers such that $\alpha + \beta = \frac{a}{a+b}$, then $[r\beta, r\alpha, \alpha, \beta]$ is a maximin strategy for this game.

ii) Show that if $b^2 = ac$ then every maximin strategy has the form specified in part i.

iii) Show that if $b^2 > ac$ and $\frac{b}{a} \geq r \geq \frac{c}{b}$, then $[0, \frac{r}{1+r}, \frac{1}{1+r}, 0]$ is a maximin strategy for this game.

iv) Show that if $b^2 > ac$ then every maximin strategy has the form specified in part iii.

v) Show that if $b^2 < ac$ and $\frac{b}{a} \leq r \leq \frac{c}{b}$, then $\left[\frac{r}{1+r}, 0, 0, \frac{1}{1+r}\right]$ is a maximin strategy for this game.

vi) Show that if $b^2 < ac$ then every maximin strategy has the form specified in part v.

15*. Is Penny-matching a symmetric game?

16*. Prove that every symmetric 2×2 game is strictly determined.

17*. For what value of a is the game

1	0	a
0	1	0
0	0	1

fair? Find the corresponding maximin and minimax strategies.

18*. For what values of a, b is the game

1	0	$-a$
0	1	0
$-b$	0	1

fair? Find the corresponding maximin and minimax strategies.

10

POKER-LIKE GAMES

Some very simple poker-like games are modeled and solved as zero-sum games.

At first glance, many popular games, such as Poker and Monopoly, even when they are played by only two players, seem very different from the games that have been discussed so far. While these former games do constitute situations of conflict, and they are zero-sum in that one player's gain constitutes the other player's loss, the nature of their moves does not appear to conform to the format posited by the mathematical theory of games. The latter requires that the players exercise their options simultaneously and independently of each other, whereas the players of Poker and Monopoly alternate their moves and are fully aware of each other's actions.

This apparent shortcoming of the theory can be remedied by a careful interpretation of the meaning of a player's options, in other words, his pure strategies. Here we return to the original, nonmathematical, sense of the word *strategy* as a guiding principle for making specific decisions. However, an accurate mathematical analysis requires more precise and detailed guidelines than such vague prescriptions as "Call if your opponent is holding his breath" or "Bluff if the moon is full." For us here a pure strategy is a set of unequivocal instructions which tells the player exactly what to do in any set of circumstances. We shall assume that each player begins each play by first settling in his mind on some such strategy for that play. He will of course give his opponent no hint about his mindset, and he may (and probably should) change his strategy from one play to another. This changing of strategies is of course tantamount to the use of a mixed strategy in the sense of the previous chapters.

In Poker (or Monopoly) a single such detailed set of instructions would fill many pages, and the number of such instruction booklets (i.e., the number of pure strategies) that the theory would have to take into account is literally unimaginable. For these reasons games like Poker and Monopoly have so far defied mathematical analysis. Nevertheless, it is possible to describe and solve some highly simplified versions of Poker where the pure strategies are completely specifiable and where the mathematical solution does provide some food for thought.

EXAMPLE 1. Two cards marked H (for high) and L (for low) are placed in a hat. Ruth draws a card and inspects it. She may then *fold* in which case she pays Charlie amount a, or she may *bet* in which case Charlie may either *fold*, paying amount a to Ruth, or *call*. If Charlie *calls* he receives from or pays to Ruth amount b according as Ruth's card is marked H or L.

Figure 1 contains a diagram illustrating the players' options after Ruth draws her card. As such diagrams describe a branching process they are called *game trees*.

Since Charlie does not know which card is drawn, when it comes to his turn he has only two options which are in fact also his pure strategies. He can either fold or call. Ruth, on the other hand, can take her card into account when deciding what to do. In principle she therefore has four pure strategies:

$$\begin{matrix} Fold \\ Fold \end{matrix} \quad i.e. \quad \begin{matrix} Fold \text{ if card } = H \\ Fold \text{ if card } = L \end{matrix}$$

$$\begin{matrix} Fold \\ Bet \end{matrix} \quad i.e. \quad \begin{matrix} Fold \text{ if card } = H \\ Bet \ \text{ if card } = L \end{matrix}$$

$$\begin{matrix} Bet \\ Fold \end{matrix} \quad i.e. \quad \begin{matrix} Bet \ \text{ if card } = H \\ Fold \text{ if card } = L \end{matrix}$$

$$\begin{matrix} Bet \\ Bet \end{matrix} \quad i.e. \quad \begin{matrix} Bet \text{ if card } = H \\ Bet \text{ if card } = L \end{matrix}$$

The actual payoffs of course depend on the nature of Ruth's card. They are exhibited, as Ruth's gains and losses, first in Figure 2 as two modifications of Figure 1, and then in Table 1, in the usual array form. In these arrays, Ruth's actual action (which depends on the card in her hand) is underlined. Some of these entries also call for clarification. For instance, it could be argued that since Ruth's folding terminates the play, it does not make any sense to insert any entry into the top row of either of these arrays. After all, when Ruth folds Charlie never gets a chance to either *fold* or *call*. We counter this by reiterating that Ruth's $\begin{smallmatrix} Fold \\ Fold \end{smallmatrix}$ and Charlie's *Fold* do not denote actual actions. Instead, they denote strategies that the players can decide (before the play) to adopt. Thus, the $-a$ in the upper left hand of each of the arrays of Table 1 denotes the fact that if Ruth adopts the pure strategy of folding (no matter what) and if Charlie adopts the pure strategy of folding on his turn, then each play will result in Ruth losing amount a, even though Charlie will never get a chance to actually fold.

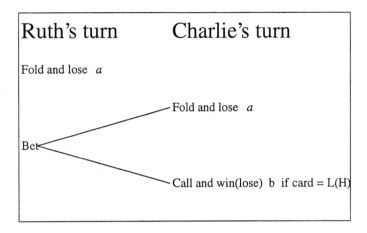

FIGURE 10.1. A Simplified Poker.

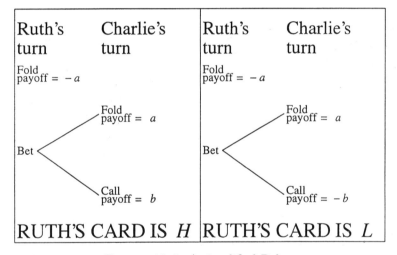

FIGURE 10.2. A simplified Poker.

TABLE 10.1. Arrays for a Simplified Poker.

	Fold	Call
Fold / Fold	$-a$	$-a$
Fold / Bet	$-a$	$-a$
Bet / Fold	a	b
Bet / Bet	a	b

RUTH'S CARD IS H

	Fold	Call
Fold / Fold	$-a$	$-a$
Fold / Bet	a	$-b$
Bet / Fold	$-a$	$-a$
Bet / Bet	a	$-b$

RUTH'S CARD IS L

Inasmuch as Ruth's chances of getting the H card are 50% and the same goes for the L card, it follows that Ruth can expect the payoff from each of the arrays of Table 1 50% of the time. Consequently these two tables are summarized into one (Table 2) by averaging the corresponding payoffs. The indicated dominance patterns are a consequence of the fact that a and b are understood to be positive numbers. Taking these dominances into account we obtain the subgame below.

TABLE 10.2. Simplified Poker as a zero-sum game.

Fold Fold	$-a$	$-a$
Fold Bet	0	$(-a-b)/2$
Bet Fold	0	$(b-a)/2$
Bet Bet	a	0

	Fold	Call
Bet Fold	0	$\frac{b-a}{2}$
Bet Bet	a	0

The solution of this subgame calls for distinguishing two cases.

CASE 1. $b \le a$. In this case the entry $\frac{b-a}{2}$ is either negative or 0, and so the lower right hand entry is a saddle point. The value of this game is 0, Ruth's pure maximin strategy is $\frac{Bet}{Bet}$ (i.e., bet on any card) and Charlie's pure minimax strategy is always *Call* on his turn.

CASE 2. $b > a$. This game is not strictly determined. Consequently, the maximin and minimax strategies are mixed. In other words these strategies dictate that in contrast with Case 1, Ruth should sometimes fold when holding an L and that occasionally Charlie should fold on his turn. For example, if $a = 4$ and $b = 6$, then the subgame is

	Fold	Call
Bet Fold	0	1
Bet Bet	4	0

Here Ruth's oddments are [4,1] and her maximin strategy is [.8,.2], i.e., she should bluff (bet on a low card) 20% of the time. Charlie's oddments are [1, 4], resulting in a minimax strategy of [.2,.8]. The value of the game is $.8 \times 0 + .2 \times 4 = .8$. The general solution of Case 2 is relegated to Exercise 1.

EXAMPLE 2. Each player places an amount a in the pot. In each of two hats two cards marked H and L are placed (so that there is a total of four cards).

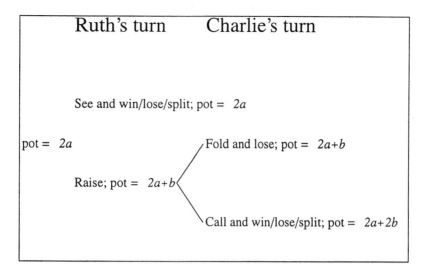

FIGURE 10.3. A simplified Poker.

Each player draws a card at random from his designated hat. Ruth has two alternatives now: she can *see* (i.e., challenge Charlie) or she can *raise* by adding amount b to the pot. If she sees, the higher hand wins the pot and equal hands split it. If she raises, Charlie has two options: he can *fold* or he can also add amount b to the pot and *call*. If he folds, Ruth wins the pot. If he calls then again the higher hand wins and equal hands split the pot. These are all the rules and they are summarized in Figure 3.

Ruth has the following four pure strategies available to her

See	See	Raise	Raise
See	Raise	See	Raise

where the top entry refers to the preferred action if Ruth draws an H and the bottom entry to the preferred action if she draws an L. Thus, the first of these calls for her seeing regardless of the nature of the card in her hand whereas the third strategy $\frac{Raise}{See}$ describes a decision to raise if the card is H and see if the card is L.

Charlie also has four pure strategies available which we denote as:

Fold	Fold	Raise	Raise
Fold	Raise	Fold	Raise

As before, the top(bottom) entry describes the action to be taken if Charlie's card is $H(L)$.

The payoffs depend on the cards that are drawn. There are four possible ways the cards can be drawn and the four corresponding payoff arrays are described in Figures 4-7. In each of the pure strategies the action taken by the player is underlined.

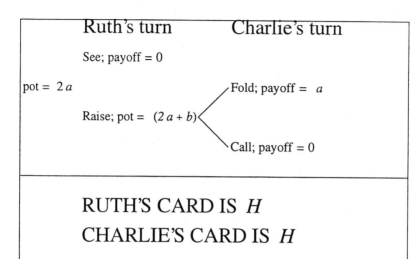

	Fold Fold	Fold Call	Call Fold	Call Call
See See	0	0	0	0
See Raise	0	0	0	0
Raise See	a	a	0	0
Raise Raise	a	a	0	0

FIGURE 10.4. A simplified Poker.

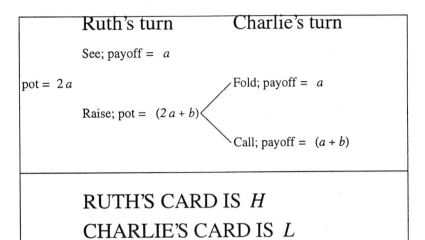

	Fold Fold	Fold Call	Call Fold	Call Call
See See	a	a	a	a
See Raise	a	a	a	a
Raise See	a	$a+b$	a	$a+b$
Raise Raise	a	$a+b$	a	$a+b$

FIGURE 10.5. A simplified Poker.

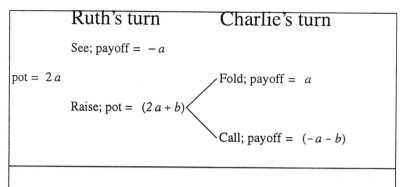

	Fold Fold	Fold Call	Call Fold	Call Call
See See	$-a$	$-a$	$-a$	$-a$
See Raise	a	a	$-a-b$	$-a-b$
Raise See	$-a$	$-a$	$-a$	$-a$
Raise Raise	a	a	$-a-b$	$-a-b$

FIGURE 10.6. A simplified Poker.

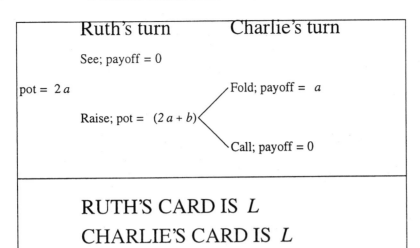

	Fold Fold	Fold Call	Call Fold	Call Call
See See	0	0	0	0
See Raise	a	0	a	0
Raise See	0	0	0	0
Raise Raise	a	0	a	0

FIGURE 10.7. A simplified Poker.

Since the four possible card distributions occur with equal probability, the four arrays of Figures 4–7 are combined into a single zero-sum game by averaging the corresponding entries (Table 3).

The dominance patterns indicated in Table 3 are a consequence of the fact that a and b are positive quantities. The resulting subgame is depicted in Table 4.

TABLE 10.3. A Simplified Poker as a Zero-sum Game.

	Fold Fold	Fold Call	Call Fold	Call Call
See See	0	0	0	0
See Raise	$3a/4$	$a/2$	$(a-b)/4$	$-b/4$
Raise See	$a/4$	$(a+b)/4$	0	$b/4$
Raise Raise	a	$(3a+b)/4$	$(a-b)/4$	0

TABLE 10.4. A Simplified Poker as a Reduced Zero-sum Ggame.

	Call Fold	Call Call
Raise See	0	$\frac{b}{4}$
Raise Raise	$\frac{a-b}{4}$	0

Observe that our analysis so far indicates that Ruth should always raise if her card is H and Charlie should always call if his card is H. The solution of the subgame depends on the values of a and b in a somewhat complicated manner and we discuss only two special cases, leaving the remainder to Exercises 6, 7. Note that the two strategies $\frac{Raise}{See}$ (for Ruth) and $\frac{Call}{Fold}$ (for Charlie) suggest that the respective player act cautiously if his card is L, whereas the two remaining strategies $\frac{Raise}{Raise}$ (for Ruth) and $\frac{Call}{Call}$ (for Charlie) dictate an aggressive reaction to that card. Accordingly, the first two will be described as *conservative* whereas the last two as *bluffing*.

If $a = 4$ and $b = 8$ the subgame becomes

		Call Fold Conservative	Call Call Bluffing
Raise See	Conservative	0	2
Raise Raise	Bluffing	-1	0

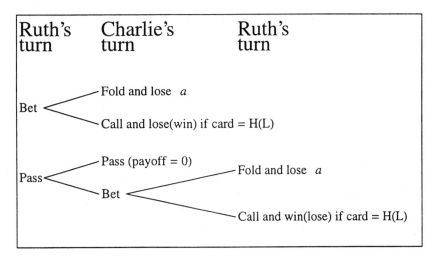

Ruth's turn | Charlie's turn | Ruth's turn

Bet — Fold and lose *a*
Bet — Call and lose(win) if card = H(L)

Pass — Pass (payoff = 0)
Pass — Bet — Fold and lose *a*
Bet — Call and win(lose) if card = H(L)

FIGURE 10.8. A simplified Poker.

which has a saddle point with value 0 in the upper left hand corner and which suggests that both players consistently play conservatively.

If $a = 8$ and $b = 4$ the subgame becomes

		Call Fold Conservative	Call Call Bluffing
Raise See	Conservative	0	1
Raise Raise	Bluffing	1	0

This game is not strictly determined. The subgame's maximin strategy of $[.5, .5]$ suggests that upon drawing an L Ruth should bluff 50% of the time. A similar bluffing policy is suggested by the minimax strategy of $[.5, .5]$. The value of the game is .5.

EXAMPLE 3. Two cards marked H and L are placed in a hat. Ruth picks a card at random and inspects it. She then may either *bet* or *pass*. If Ruth bets, Charlie may either *fold* and pay a, or else he can *call* and either lose or win amount b according as Ruth's card is H or L. If Ruth passes, Charlie may decide to *pass*, in which case the payoff to each player is 0, or he may *bet*, in which case the play reverts to Ruth again. This time she may *fold* and lose a, or she may *call* winning or losing amount b according as her card is or is not H.

The rules of this game are summarized in Figure 8.

Given any card in her hand, Ruth may decide to either bet or pass. If she passes she has to be ready to face the contingency that Charlie will choose to bet and then she will have to either fold or call. Thus, Ruth needs to take three move sequences into account when planning her strategy:

B – Bet.

PF – Pass and fold if Charlie bets.

PC – Pass and call if Charlie bets.

Consequently, Ruth has 9 pure strategies:

B	B	B	PF	PF	PF	PC	PC	PC
B	PF	PC	B	PF	PC	B	PF	PC

In these strategies the top entry specifies the move sequence to be taken if Ruth gets card H, and the bottom entry specifies the action to be taken if she gets card L.

On the other hand, Charlie has four pure strategies:

$\begin{matrix} F \\ P \end{matrix}$ *i.e.*   Fold if Ruth bets / Pass if Ruth passes

$\begin{matrix} F \\ B \end{matrix}$ *i.e.*   Fold if Ruth bets / Bet if Ruth passes

$\begin{matrix} C \\ P \end{matrix}$ *i.e.*   Call if Ruth bets / Pass if Ruth passes

$\begin{matrix} C \\ B \end{matrix}$ *i.e.*   Call if Ruth bets / Bet if Ruth passes

The game trees and payoff arrays that correspond to cards H and L are displayed in Figures 9 and 10 respectively.

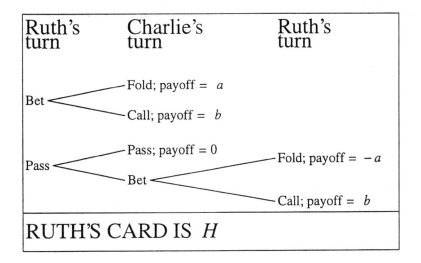

	F P	F B	C P	C B
$\frac{B}{B}$	a	a	b	b
$\frac{B}{PF}$	a	a	b	b
$\frac{B}{PC}$	a	a	b	b
$\frac{PF}{B}$	0	$-a$	0	$-a$
$\frac{PF}{PF}$	0	$-a$	0	$-a$
$\frac{PF}{PC}$	0	$-a$	0	$-a$
$\frac{PC}{B}$	0	b	0	b
$\frac{PC}{PF}$	0	b	0	b
$\frac{PC}{PC}$	0	b	0	b

FIGURE 10.9. A simplified Poker.

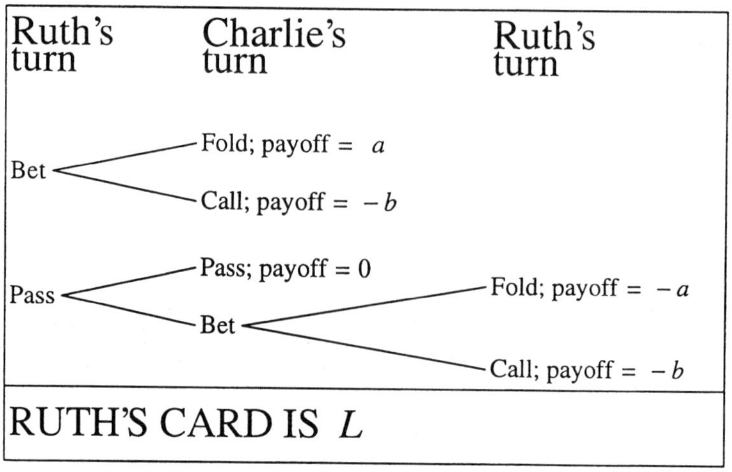

Ruth's turn	Charlie's turn	Ruth's turn
	Fold; payoff = a	
Bet <		
	Call; payoff = $-b$	
	Pass; payoff = 0	
Pass <		Fold; payoff = $-a$
	Bet <	
		Call; payoff = $-b$

RUTH'S CARD IS L

	F P	F B	C P	C B
B B̲	a	a	$-b$	$-b$
B P̲F̲	0	$-a$	0	$-a$
B P̲C̲	0	$-b$	0	$-b$
PF B̲	a	a	$-b$	$-b$
PF P̲F̲	0	$-a$	0	$-a$
PF P̲C̲	0	$-b$	0	$-b$
PC B̲	a	a	$-b$	$-b$
PC P̲F̲	0	$-a$	0	$-a$
PC P̲C̲	0	$-b$	0	$-b$

FIGURE 10.10. A simplified Poker.

Since the probability of Ruth getting any specific card is 50%, the two tables of Figures 9 and 10 are summarized by averaging their respective entries. The resulting array is displayed in Table 5.

While the general solution of this game lies beyond the scope of this text, enough tools were provided in the previous chapters to enable us to draw interesting conclusions in some special cases.

If $a \geq b$ then the game of Table 5 has a saddlepoint in its upper right hand entry. Thus, this game is strictly determined. Ruth's maximin strategy calls for consistent betting, regardless of her card, and Charlie minimax response is to consistently call her bets. The value of this game is 0.

TABLE 10.5. A Simplified Poker as a
Zero-sum Game

	F P	F B	C P	C B
B B	a	a	0	0
B PF	$\frac{a}{2}$	0	$\frac{b}{2}$	$\frac{b-a}{2}$
B PC	$\frac{a}{2}$	$\frac{a-b}{2}$	$\frac{b}{2}$	0
PF B	$\frac{a}{2}$	0	$\frac{-b}{2}$	$\frac{-a-b}{2}$
PF PF	0	$-a$	0	$-a$
PF PC	0	$\frac{-a-b}{2}$	0	$\frac{-a-b}{2}$
PC B	$\frac{a}{2}$	$\frac{a+b}{2}$	$\frac{-b}{2}$	0
PC PF	0	$\frac{b-a}{2}$	0	$\frac{b-a}{2}$
PC PC	0	0	0	0

Suppose, on the other hand, that $a = 2$ and $b = 4$, in which case the game reduces to

TABLE 10.6. A Reduced Simplified Poker.

	F P	F B	C P	C B
B B	2	2	0	0
B PF	1	0	2	1
PC B	1	3	-2	0
PC PF	0	1	0	1

Techniques that lie outside the scope of this text can be used to derive the following solution to this game (see Exercises 20, 21, 22, though).

$$\text{value} = .8$$
$$\text{maximin strategy} = [.2, .4, 0, .4]$$
$$\text{minimax strategy} = [.2, .2, 0, .6].$$

It is noteworthy that this maximin strategy suggests that even when Ruth holds a high card she should pass 40% of the time, rather than cash the high card in immediately. It is also interesting that while the pure strategy $\begin{smallmatrix} C \\ P \end{smallmatrix}$ looks at first very promising for Charlie since only one of its entries is positive and the sum of its entries is 0, the minimax strategy specifies that this pure strategy should never be used by Charlie.

Chapter Summary

Three very simple variations on the game of poker are modeled as zero-sum games and then solved.

Chapter Terms

Game tree   98   Strategy   97

Exercises 10

In all the following exercises the quantities $a, b, c, \ldots$ are assumed to be positive.

1. Show that the general solution to the subgame of Table 10.2 when $b > a$ is value $= \frac{a(b-a)}{a+b}$, maximin strategy $= \left[\frac{2a}{a+b}, \frac{b-a}{a+b}\right]$, minimax strategy $= \left[\frac{b-a}{a+b}, \frac{2a}{a+b}\right]$.

2. Suppose that in the game of Example 1, Charlie pays amount b (rather than a) when he folds. Show that this leads to the subgame

$\frac{b-a}{2}$	$\frac{b-a}{2}$
b	0

and solve this subgame. (Hint: consider the two cases $b \le a$ and $b > a$ separately.)

3. Suppose that in the game of Example 1, Ruth loses amount b (rather than a) when she folds. Show that this game is strictly determined and solve it.

4. Suppose that in the game of Example 1 Charlie loses amount c (rather than a) when he folds. Show that this leads to the subgame

$\frac{c-a}{2}$	$\frac{b-a}{2}$
c	0

and solve this subgame.

5. Suppose that in the game of Example 1 Charlie loses amount a (rather than b) if he calls and Ruth turns out to be holding H. Show that this game is strictly determined and solve it.

6. Show that if $a > b$ then the game of Table 10.3 is not strictly determined. Show that in this case the game has value $\frac{b(a-b)}{4a}$ and its maximin and minimax strategies are $\left[0, 0, \frac{a-b}{a}, \frac{b}{a}\right]$ and $\left[0, 0, \frac{b}{a}, \frac{a-b}{b}\right]$ respectively.

7. Show that if $a \leq b$ then the game of Table 10.3 is strictly determined and solve it.

8. Suppose that in the game of Example 2 Charlie adds amount a (rather than b) to the pot. Show that the resulting game can be reduced to the subgame

0	0
$\frac{a-b}{4}$	$\frac{a-3b}{8}$
$\frac{a-b}{8}$	$\frac{3a-b}{8}$
$\frac{3a-3b}{8}$	$\frac{a-b}{2}$

9. Solve the game of Exercise 8 if $a = 16$ and $b = 8$.

10. Solve the game of Exercise 8 if $a = 8$ and $b = 16$.

11. Solve the game of Exercise 8 for any positive a and b.

12. Suppose that in the game of Example 2 Ruth adds amount a (rather than b) to the pot when she raises. Show that the resulting game can be reduced to the subgame

0	0	0
$\frac{3a}{4}$	0	$\frac{b-3a}{8}$
$\frac{a}{4}$	$\frac{b-a}{8}$	$\frac{3b-a}{8}$
a	$\frac{b-a}{8}$	$\frac{b-a}{2}$

13. Solve the game of Exercise 12 if $a = 8$ and $b = 16$.

14. Solve the game of Exercise 12 if $a = 16$ and $b = 8$.

15. Solve the game of Exercise 12 if $a \geq b$.

16. Solve the game of Exercise 12 if $b \geq 3a$.

17. Show that if $a \geq b$ in the game of Example 3, then the game is strictly determined and solve it.

18. Suppose that in the game of Example 3, Ruth loses amount b (rather than a) whenever she folds. Show that the resulting game is strictly determined and solve it.

19. Suppose that in the game of Example 3, Charlie loses amount b (rather than a) whenever he folds. Show that the resulting game is strictly determined and solve it.

20. Show that if Ruth employs strategy $[.2, .4, 0, .4]$ in the subgame of Table 10.6 and Charlie employs any pure strategy, then the resulting payoff is no less than .8.

21. Show that if Charlie employs strategy $[.2, .2, 0, .6]$ in the subgame of Table 10.6 and Ruth employs any pure strategy, then the expected payoff is no greater than .8.

22. Use Exercises 21, 22 to argue that value $= .8$, maximin strategy $= [.2, .4, 0, .4]$ and minimax strategy $= [.2, .2, 0, .6]$ constitute a solution to the subgame of Table 10.6.

23. Show that if $b \geq a$ then Example 3 reduces to the subgame

	F P	F B	C P	C B
B B	a	a	0	0
B PF	$\frac{a}{2}$	0	$\frac{b}{2}$	$\frac{b-a}{2}$
PC B	$\frac{a}{2}$	$\frac{a+b}{2}$	$\frac{-b}{2}$	0
PC PF	0	$\frac{b-a}{2}$	0	$\frac{b-a}{2}$

Show that

$$\text{value} = \frac{ab(b-a)}{b^2 + a^2}$$

$$\text{maximin strategy} = \left[\frac{(b-a)^2}{b^2 + a^2}, \frac{2a(b-a)}{b^2 + a^2}, 0, \frac{2a^2}{b^2 + a^2} \right]$$

$$\text{minimax strategy} = \left[\frac{(b-a)^2}{b^2 + a^2}, \frac{a(b-a)}{b^2 + a^2}, 0, \frac{a(b+a)}{b^2 + a^2} \right]$$

constitutes a solution to this subgame.

24. Solve the following variation on the game of Example 2. Each player places an amount a in the pot. In each of two hats an Ace, a King and a Queen are placed (so that there is a total of six cards). Each player draws a card at random from his designated hat. Ruth has two alternatives now: she can *see* (i.e., challenge Charlie) or she can *raise* by adding amount b to the pot. If she sees, the higher hand wins the pot and equal hands split it. If she raises, Charlie has two options: he can *fold* or he can also add amount b to the pot and *call*. If Charlie folds, Ruth wins the pot. If he calls then again the higher hand wins and equal hands split the pot. (Hint: It clearly doesn't make any sense to adopt a strategy that simultaneously calls for an aggressive move on one card and a conservative move on a higher card. Consequently each player needs consider only three strategy: aggressive on all three cards, aggressive on the Ace and King only, or aggressive on the Ace only.)

11

PURE MAXIMIN AND
MINIMAX STRATEGIES

What to do when a zero-sum game is played only once.

Sometimes a game such as Bombing-sorties is played only one time, so that

Attack

		Bomber	Support
Bomb placement	Bomber	80%	100%
	Support	90%	50%

the von Neumann Theorem is not applicable. Nevertheless, the basic idea of the maximin and minimax strategies can still be used to make decisions in a manner that seems to correspond quite realistically to the manner in which people actually think. Note that in this game, if General Ruth places the bomb on the bomber she is guaranteed a payoff of an 80% chance of the mission being successful. On the other hand, the alternative of placing the bomb on the support plane can only guarantee a 50% chance of success. The U. S. military directives are quite explicit on this subject. *A commander is enjoined to select that course of action which offers the greatest promise of success in view of all the enemy capabilities.* In other words, that course of action is to be pursued which can guarantee, regardless of enemy actions, the best possible results. In this case, the directive calls for placing the bomb on the bomber, since that guarantees a payoff of 80%, regardless of the fact that placing the bomb on the support plane *might* result in the higher 90% if Charlie attacks the bomber (as he most likely will). This conservative search for guarantees is applicable to any zero-sum game. For Ruth this is called the *pure maximin strategy* and it dictates the selection of that

row whose minimum entry is the largest (amongst all the rows). That largest
minimum entry is called the *pure maximin value* of the game.

EXAMPLE 1. Find the pure maximin strategy and value of the game

3	2	6	2
5	4	3	4
1	2	3	1

Each row's minimum entry is tallied on the right and the largest of these is 3.

3	2	6	2	2
5	4	3	4	<u>3</u>
1	2	3	1	1

Consequently, this game's pure maximin strategy is $[0, 1, 0]$ and its pure maximin
value is 3.

In Bombing-sorties, Charlie's military directives imply that he should attack
the bomber, since that reduces the mission's chance of success to 80%. He
would be foolish to attack the support plane on the offchance that it might
hold the bomb, since that might allow for a 100% chance of success if the bomb
is on the other plane. In fact, General Charlie must choose the column that
guarantees the least chance of success for Ruth, *no matter what Ruth does!* This
is formalized as the *pure minimax strategy* of an arbitrary zero-sum game, which
calls for Charlie's selection of that column whose maximum entry is the smallest
(amongst all the columns). That smallest maximum entry is the *pure minimax
value* of the game.

EXAMPLE 2. To find a pure minimax strategy and the value of the game of
the previous example, each column's maximum entry is tallied at the bottom.

3	2	6	2
5	4	3	4
1	2	3	1
5	<u>4</u>	6	<u>4</u>

Since the smallest of these maxima is 4, it follows that both $[0, 1, 0, 0]$ and $[0,
0, 0, 1]$ constitute pure minimax strategies and the pure minimax value is 4.

For any single instance of the game of Examples 1, 2, Ruth can guarantee,
by selecting the second row, that she will gain at least 3. Charlie, on the other

hand, can guarantee, by selecting the second column, that Ruth will gain no more than 4. Since this is true for any single instance even when the game is played many times, it follows that the (mixed) value of the game must lie between these guarantees. In general the following holds.

THEOREM 3. *For any $m \times n$ zero-sum game*

$$\text{pure maximin value} \leq \text{mixed value} \leq \text{pure minimax value.}$$

EXAMPLE 4. In the game of Examples 1,2 the third row is dominated by the first one and may consequently be deleted. In the resulting 2×4 game the first and the last columns both dominate the second one and can be deleted. The resulting subgame

2	6
4	3

is not strictly determined. Ruth has oddments $[4-3, 6-2] = [1, 4]$ and so her maximin strategy is $[.2, .8]$. The value of both the subgame and the original game is

$$.2 \times 1 \times 2 + .8 \times 1 \times 4 = .4 + 3.2 = 3.6$$

which does indeed lie between the pure maximin value of 3 and the pure minimax value of 4.

Chapter Summary

When a zero-sum game is played only once the mixed minimax and maximin strategies are of no avail. Instead pure maximin and minimax strategies (and values) are recommended. The relation between the pure and mixed solutions of a game is displayed in Theorem 3.

Chapter Terms

Pure minimax strategy	116	Pure minimax value	116
Pure maximin strategy	115	Pure maximin value	116

EXERCISES 11

Find the pure maximin and minimax strategies and values for the games in Exercises 1–8.

1.

2	−1
0	2

2.

2	−1	0
0	2	−1

3.

0	−3
1	2
−1	5

4.

1	2	3
4	5	6
7	8	9

1	9	1	5
6	2	6	0
5	7	3	2
8	4	3	4

5.

0	−1	−2	3
−2	1	0	5
4	−2	2	1

6.

7.

−2	3	−1	−4	9
4	−9	−5	0	2
5	−5	6	−6	0
5	3	1	2	4

8.

−9	1	2	3	−4
4	−8	5	−3	6
7	0	−7	0	8
1	−2	3	−6	3
−1	4	5	6	−5

Verify the statement of Theorem 3 for each of the games in Exercises 9–14.

9.

2	−1
0	2

10.

−1	5
0	−2

11.

3	4
0	−5

12.

2	3
0	9

13.

5	−1
−2	3

14.

2	0
−1	2

15*. Prove that for any 2 × 2 game, if any two of the following three quantities are equal to each other then all three are equal to each other: *the value of the game, the pure maximin value of the game, the pure minimax value of the game.* Find a 2 × 3 game for which this is false.

16. The payoffs in the game below denote dollars. Ruth and Charlie are about to play this game 100 times.

3	−1
1	2

    a) Does Ruth have a strategy that guarantees her winning something each time the game is played?

    b) Does Ruth have a strategy that guarantees her winning a total of $80 or more?

    c) Does Ruth have a strategy that guarantees her winning a total of $120 or more?

    d) Ruth is playing this game in order to raise $120 which she needs very badly (she owes this money to a loan shark). What strategy should she use?

    e) Does Charlie have a strategy that is guaranteed to hold Ruth's winnings to less than $90?

    f) Does Charlie have a strategy that is guaranteed to hold Ruth's winnings to less than $120?

    g) It so happens that Charlie only has $125. Should he lose more, he will have to ask his parents again for some money, something he is not willing to do anymore. If he still wants to play, what should his strategy be?

17. The payoffs in the game below denote dollars. Ruth and Charlie are about to play this game 100 times.

3	−2
−1	2

    a) Does Ruth have a strategy that guarantees her winning something each time the game is played?

    b) Does Ruth have a strategy that guarantees that her losses will not exceed $120?

    c) Does Ruth have a strategy that guarantees that her losses will not exceed $80?

    d) Ruth is playing this game in order to raise $30 which she needs very badly (she owes this money to a loan shark). What strategy should she use?

    e) Does Charlie have a strategy that is guaranteed to hold Ruth's winnings to less than $90?

    f) Does Charlie have a strategy that is guaranteed to hold Ruth's winnings to less than $120?

    g) It so happens that Charlie only has $60. Should he lose more, he will have to ask his parents again for some money, something he is not willing to do anymore. If he still wants to play, what should his strategy be?

18. The payoffs in the game below denote dollars. Ruth and Charlie are about to play this game 100 times.

3	2
−1	2

    a) Does Ruth have a strategy that guarantees her winning something each time the game is played?

    b) Does Ruth have a strategy that guarantees her winning a total of $200 or more ?

    c) Does Ruth have a strategy that guarantees her winning a total of $300 or more?

    d) Ruth is playing this game in order to raise $30 which she needs very badly (she owes this money to a loan shark). What strategy should she use?

    e) Does Charlie have a strategy that is guaranteed to hold Ruth's winnings to less than $90?

    f) Does Charlie have a strategy that is guaranteed to hold Ruth's winnings to less than $250?

    g) It so happens that Charlie only has $250. Should he lose more, he will have to ask his parents again for some money, something he is not willing to do anymore. If he still wants to play, what should his strategy be?

19. The payoffs in the game below denote dollars. Ruth and Charlie are about to play this game 100 times.

3	−3
−4	2

    a) Does Ruth have a strategy that guarantees her winning something each time the game is played?

    b) Does Ruth have a strategy that guarantees that her losses will not exceed $120?

    c) Does Ruth have a strategy that guarantees that her losses will not exceed $300?

    d) Charlie is playing this game in order to raise $30 which he needs very badly (he owes this money to a loan shark). What strategy should he use?

    e) Does Charlie have a strategy that is guaranteed to hold Ruth's winnings to less than $180?

    f) Does Charlie have a strategy that is guaranteed to hold Ruth's winnings to less than $250?

    g) It so happens that Ruth only has $35. Should she lose more, she will have to ask her parents again for some money, something she is not willing to do anymore. If she still wants to play, what should her strategy be?

20*. Prove Theorem 3.

21*. Suppose the rules of game playing are changed so that Charlie is allowed to make his move after Ruth has made her move and in full knowledge of her actions. Identify Ruth's optimal strategy.

22*. Suppose the rules of game playing are changed so that Ruth is allowed to make her move after Charlie has made his move and in full knowledge of his actions. Identify Charlie's optimal strategy.

PURE NONZERO-SUM GAMES

Some well known concrete nonzero-sum games are discussed.
Solutions are offered for the unrepeated games and
their characteristics are appraised.

A considerable amount of scholarly effort has been expended on attempts to carry over parts of the theory of zero-sum games to other, less restrictive, games. These are games wherein one player's gain is not necessarily the other player's loss, where the payoff may be unquantifiable, and where the players' decisions may not be independent of each other. Sometimes the payoff may be oneupmanship or loss of face, and in some situations reasonable behavior on the part of the players will lead to greater payoffs for both of them. Of the many generalizations of zero-sum games, the remainder of this text is confined to a discussion of *two-person noncooperative nonzero-sum games*. The term *noncooperative* refers to the assumption that the players do not consult each other about ways and means for improving their payoffs. In the interest of brevity these games are referred to as simply *nonzero-sum games* in the sequel.

Nonzero-sum games are also modeled by having Ruth select a row and Charlie select a column from a suitable rectangular array. However, since a player's gain is not necessarily his opponent's loss anymore, the payoff of each outcome can no longer be described by a single number. The result of Ruth choosing row i and Charlie choosing column j is the *payoff pair*

$$(a_i, b_j)$$

where a_i denotes Ruth's payoff and b_j denotes Charlie's payoff. Thus, the general 2×2 nonzero-sum game has the array

Charlie

(a_1, b_1)	(a_2, b_2)
(a_3, b_3)	(a_4, b_4)

Ruth

$$(1)$$

With some care, the zero-sum game strategies can be applied here as well. First, however, it is necessary to formalize a guiding principle that presumably motivates the players decisions. In the zero-sum case Ruth looked to maximize her payoff, and, since Ruth's gain was his loss, Charlie sought to minimize the same payoff. Now, since the two players' payoffs are independent of each other, it will be assumed that each player bases his own decisions on his payoffs alone. This is commonly formalized as follows.

THE PRINCIPLE OF RATIONALITY. *Every player wishes to come out as well off as possible.*

In other words, cutting off one's nose in order to spite one's spouse is not rational behavior, and such impulses will be given no consideration in this text. A consequence of the Principle of Rationality is that each player will ignore the other player's payoffs in forming his decision. Thus, we may think of Ruth as facing the (zero-sum) game

Ruth

a_1	a_2
a_3	a_4

against an irrelevant opponent and of Charlie as facing the (zero-sum) game

Charlie

b_1	b_2
b_3	b_4

wherein his object, just like Ruth's is to *maximize* his payoff. Consequently, both players have *maximin* strategies. The payoff pair in the outcome determined by the two pure maximin solutions is called the *pure value pair* of the game. The pure maximin strategies and pure value pairs of any nonzero-sum game are easily derived.

EXAMPLE 1. Find the pure maximin strategies and pure value pairs of the game

$(2, 3)$	$(3, 5)$
$(2, 2)$	$(1, 3)$

By the Principle of Rationality, Ruth's considerations are confined to the array

Ruth

2	3	2
2	1	1

and so her pure maximin strategy is [1, 0] with a corresponding maximin value of 2. On the other hand, Charlie's analysis is confined to the array

Charlie

3	5
2	3

2   3

and so his maximin strategy is [0, 1] with a corresponding maximin value of 3. Note that the entry at the bottom of each column is the *minimum* for that column and that Charlie will choose the largest of these minima, in contrast with the pure minimax stratetgy, which calls for recording the maxima of each column and choosing the smallest amongst them.

The pure value pair is therefore the entry of the first row and second column, $(3, 5)$. Notice that this pure value pair does *not* consist of the two pure maximin values. In subsequent examples Ruth's and Charlie's payoffs will not be separated and the tallies will be kept next to the array of the original game. Thus, this example's two separate arrays for Ruth and Charlie will be replaced by the single diagram below which still contains all the required information.

(2,3)	(3,5)	2
(2,2)	(1,3)	1

2   3

EXAMPLE 2. Find the pure maximin strategies and pure value pairs of the game

(−1,0)	(1,3)
(2,2)	(0,1)

The row minima of Ruth's and Charlie's arrays are both tallied below.

$(-1,0)$	$(1,3)$
$(2,2)$	$(0,1)$

-1

$\underline{0}$

$0 \qquad \underline{1}$

Since Ruth's pure maximin value is 0 it follows that Ruth's pure maximin strategy is to select the second row. Charlie's pure maximin value is 1 and so his pure maximin strategy is also to select the second row of the b-array, i.e., the second column of the given game. The pure value pair is $(0, 1)$.

PRISONER'S DILEMMA. *Two people are arrested and held in connection with a certain robbery. The prosecution only has enough evidence to convict them of the robbery itself, but it is believed that the robbers were actually carrying guns at the time, making them liable to the more severe charge of armed robbery. The prisoners are held in separate cells and cannot communicate with each other. Each of them is offered the same deal: If you testify that your partner was armed but he does not testify against you, your sentence will be suspended while he will spend 15 years in jail. If both of you testify against each other, you will both get 10 years, and if neither of you does so, you both get 5 years in jail. Both prisoners are fully aware that both have been offered the same deal. They are given some time to think the deal over, but neither is aware of his expartner's decision.*

This game is clearly nonzero-sum and can be described as

	Refuse the deal	Accept the deal
Refuse the deal	$(-5, -5)$	$(-15, 0)$
Accept the deal	$(0, -15)$	$(-10, -10)$

where the payoff $-x$ denotes a jail term of x years. Each player's pure maximin strategy is easily derived.

$(-5, -5)$	$(-15, 0)$
$\cdot\ (0, -15)$	$(-10, -10)$

-15

$-\underline{10}$

$-15 \qquad\qquad -\underline{10}$

The pure maximin strategy is $[0, 1]$ for both players. Thus, the maximin strategy recommends that both prisoners accept the deal offered by the prosecution, thereby guaranteeing that each will receive a 10 year jail term. Both experience and experimentation testify to the fact that this is frequently the decision made by people in such circumstances. This is unsettling because if both of the players

had refused the deal, they would have each received the 5 year term, a distinctly
better sentence for both.

Let us digress in order to formalize this vague notion of "unsettling." In any
nonzero-sum game, a payoff pair (a, b) is said to be *better than* the payoff (a', b')
if either of the following conditions hold:

$$a > a' \quad \text{and} \quad b \geq b' \quad \text{or} \quad a \geq a' \quad \text{and} \quad b > b'.$$

Equivalently, (a, b) is better than (a', b') if it dominates it in the sense of Chapter
8 and they are also distinct. Thus, (3, 2) is better than each of the payoffs (1, 2),
(3, 1) and (2, 2). An outcome of a game is *Pareto optimal* if the game possesses
no outcome with a better payoff. In game (1) below, the outcomes with payoffs
(2, 3) and (1, 5) are both Pareto optimal.

$(2, 3)$	$(1, 5)$
$(2, 2)$	$(0, 3)$

On the other hand, in the game of Example 1 above, the outcome with payoff
(3, 5) is the only Pareto optimal outcome.

It is clear that when the players of a game are treated as a group, Pareto optimal outcomes are desirable. The unfortunate and tragic aspect of the Prisoners'
dilemma game is that both the strategy recommended by game theory and the
observed behavior of people result in the outcome with payoff $(-10, -10)$, which
is dominated by $(-5, -5)$ and hence is *not* Pareto optimal.

CHICKEN. This game describes a very common situation. *Two adversaries
are set on a collision course. If both persist, then a very unpleasant outcome,
sometimes mutual annihilation, is guaranteed. If only one of the players swerves
away (chickens) he loses the game. If both swerve, the result is a draw.*

Usually, the reward for winning this game is merely a sense of dominance, and
swerving only entails loss of face, both equally unquantifiable. The possibility of
mutual annihilation when both players persist is what makes this a nonzero-sum
situation. In that case neither player gains anything; in fact, quite frequently
both stand to lose their lives. We shall offer no solution and wish to discuss
only some paradoxical aspects of the game. For this purpose, the game will
be made concrete by transferring our attention to the Cuban missile crisis of
October 1962, certainly one of best known and most frightening Chicken games
ever played.

In that month the United States discovered that the USSR was building a
missile base in Cuba. As a small Soviet fleet was on its way to the island to
bring in supplies, the Americans set up a naval blockade. It is quite likely that if
these two fleets had met, the conflict would have escalated into a full fledged, and
probably nuclear, war. In the event, the Soviets blinked, their ships withdrew,
and the United States "won". It is our purpose here to discuss a curious wrinkle
on this confrontation.

While it is impossible to realistically quantify the various outcomes of this game, it is possible to introduce some numbers into it by a preferential ranking of the various outcomes. There were four possible outcomes for the Cuban missile crisis:

(US swerves, USSR swerves),   (US persists, USSR swerves)

(US swerves, USSR persists),   (US persists, USSR persists).

The players would rank these outcomes preferentially as follows (1 denoting the least desirable and 4 the most desirable outcomes):

US		USSR
(US persists, USSR swerves)	4	(US swerves, USSR persists)
(US swerves, USSR swerves)	3	(US swerves, USSR swerves)
(US swerves, USSR persists)	2	(US persists, USSR swerves)
(US persists, USSR persists)	1	(US persists, USSR persists).

These preferences are summarized as the nonzero-sum game:

		USSR	
		Swerve	Persist
US	Swerve	$(3,3)$	$(2,4)$
	Persist	$(4,2)$	$(1,1)$

Here the "payoff" (i,j) associated with any outcome denotes the fact that the US assigns this outcome the ranking i in its preference list, whereas the USSR assigns it the ranking j. The maximin solution of this game recommends that both players swerve and has payoff $(3,3)$.

Suppose now that on October 27, the day before the crisis ended, President Kennedy had been notified that his staff was infiltrated by Soviet spies and that whatever decision he would make would immediately be made known to Premier Khrushchov. One would think this to be a cause for much consternation for the Americans. We will argue that the opposite is the case and that in fact:

In the game of Chicken, a player's foreknowledge of his opponent's decision works to that player's disadvantage, provided that the opponent is aware of the player's foreknowledge.

This conclusion will be drawn on the basis of the Principle of Rationality. Taking the infiltration into account, President Kennedy would have reasoned as follows:

if I decide to swerve, Khrushchov will know it and, choosing between $(3,3)$ and $(2,4)$, he will decide to persist, an outcome with payoff $(2,4)$;

if I decide to persist, Khrushchov will know it and, choosing between $(4,2)$ and $(1,1)$, he will decide to swerve, an outcome with payoff $(4,2)$.

Thus, if Kennedy swerves the outcome will be (2, 4), whereas if he persists the outcome will be (4, 2). He would therefore choose to persist, resulting in the payoff pair (4, 2) wherein the Soviets swerve.

This is, of course, exactly what transpired. It could be argued that while the USSR had in all likelihood not been privy to the American decision making process, its leaders were wiser and more cognizant of the possible disastrous consequences of political games of Chicken. After all, only 17 years had passed since the conclusion of World War II in which 20 million Russians are estimated to have perished and major portions of their land were devastated whereas only one of that war's battles was fought on American soil.

The essential feature of the infiltration that was hypothesized in the Cuban Missile Crisis is that the USSR did not select its option until after the US had done so. In general this prerogative may or may not be beneficial to the player who possesses it.

EXAMPLE 3. Consider the game

$(5, 2)$	$(3, 0)$	$(8, 1)$	$(2, 3)$
$(6, 3)$	$(5, 4)$	$(7, 4)$	$(1, 1)$
$(7, 5)$	$(4, 6)$	$(6, 8)$	$(0, 2)$

Suppose first that Charlie is informed of Ruth's decision when he makes his own, and that Ruth is aware of this fact. Then Ruth will reason as follows:

If I select row 1,   Charlie will opt for the payoff pair (2, 3);
If I select row 2,   Charlie will opt for one of the payoff pairs (5, 4) and (7, 4);
If I select row 3,   Charlie will opt for the payoff pair (6, 8).

The Principle of Rationality then will direct Ruth to select row 3 and Charlie to choose column 3 resulting in an outcome with payoff pair (6, 8).

On the other hand, if Ruth is the one who is informed of Charlie's decision when she makes her own, and if Charlie is aware of this fact, then he will reason as follows:

If I select column 1,   Ruth will opt for payoff pair (7, 5);
If I select column 2,   Ruth will opt for payoff pair (5, 4);
If I select column 3,   Ruth will opt for payoff pair (8, 1);
If I select column 4,   Ruth will opt for payoff pair (2, 3).

The Principle of Rationality then will direct Charlie to choose column 1 and Ruth to choose row 3 for a payoff pair of (7, 5).

If neither player is informed of their opponent's decisions, then Ruth's maximin strategy calls for the selection of row 1 for a guaranteed payoff of at least 2. Charlie's maximin strategy calls for the selection of column 1 for a guaranteed payoff of at least 2. The actual outcome will then have payoff pair (5, 2). Thus, in this game, information about the opponent's decisions is advantageous to each player.

It was seen in the discussion of the Chicken game that the maximin analysis can fail to properly describe human behavior in some gamelike situations. In retrospect, in that particular case, the failure was due to the oversimplification that occurred when it was assumed that the players made their decision independently and simultaneously. In most instances of this game however, the players' behaviors very much affect each other. They continuously observe each other and when one swerves the other will most certainly persist to the end. Rather than completely disqualify the maximin strategy for such games, a way was found to adapt it so as to continue to supply researchers with satisfactory solutions.

Both the pure and the mixed maximin solutions of the zero-sum games possess a qualitative trait that can be transferred to the nonzero-sum context as well. Recall that both yield a maximal guaranteed payoff; an *expected* payoff in the mixed case, but still guaranteed. In other words, if the player is looking for guarantees, then he only stands to lose (i.e., have his guarantee diminished) by switching to a nonmaximin strategy. Consider, for example, the zero-sum game below in which the underlined 1 is a saddlepoint.

4	0	4
1	$\underline{1}$	1
4	0	4

The maximin strategy calls for Ruth to select the second row. Note that if she selected either of the other rows, she *might* possibly gain 4, but her *guarantee* of always winning at least 1 would be gone. Experience indicates that this search for guarantees is a very strong motivator of human behavior. While it may not lead to the best of all possible payoffs, it does offer the players the consolation of having eliminated the possibility of their regretting their decisions. Each player knows that he has played so as to obtain the best possible guarantee, regardless of what the opponent did. This *No Regrets* policy can be applied to nonzero-sum games as well.

In a nonzero-sum game, an outcome is said to be a *Nash equilibrium point* if its payoff pair (a_i, b_j) is such that a_i is maximum for its column, and b_j is maximum for its row. Such a Nash equilibrium point has the desired *No Regrets* property since, given that Charlie selected column j, Ruth has no reason to regret having selected row i—no other selection of Ruth's would have yielded her a better payoff. Similarly, given that Ruth selected row i, Charlie has no reason to regret having selected column j—no other selection of his would have yielded him a better payoff. Note that in this definition of the Nash equilibrium each player is assumed to be concerned with regrets only over her or his own actions. Any feelings that they might have regarding the other players' choices are ignored. The reason for this limitation is that the incorporation of such feelings would result in too many complications.

EXAMPLE 4. Find the Nash equilibrium points of the game

$(5,2)$	$(3,0)$	$(8,1)$	$(2,3)$
$(6,3)$	$(5,4)$	$(7,4)$	$(1,1)$
$(7,5)$	$(4,6)$	$(6,8)$	$(0,2)$

One method for finding these points is to examine all the payoff pairs (a_i, b_j) in succession and to star both each a_i that is a maximum for its column and each b_j that is a maximum for its row. The Nash equilibrium points are those

$(5,2)$	$(3,0)$	$(8^*,1)$	$(2^*,3^*)$
$(6,3)$	$(5^*,4^*)$	$(7,4^*)$	$(1,1)$
$(7^*,5)$	$(4,6)$	$(6,8^*)$	$(0,2)$

whose payoff pairs have *both* entries starred. Thus, the Nash equilibrium points of the given game are those with payoffs (5, 4) and (2, 3). Note that one of these payoff pairs, namely (5, 4), is clearly preferable to the other. Nevertheless, this does not necessarily mean that either Ruth or Charlie will aim for this "better" payoff pair. After all, by selecting the second row (that of the pair (5, 4)), Ruth opens herself up to the possibility of obtaining a payoff of only 1 if Charlie selects column 4. Similarly, if Charlie were to select the second column (that of the pair (5, 4)) he would be opening himself up to the possibility of gain 0 if Ruth selects row 1.

The next example shows that pure Nash equilibrium points need not exist.

EXAMPLE 5. Find the Nash equilibrium points of the game

$(2,1)$	$(1,2)$
$(1,2)$	$(2,1)$

When the row maximum of the first entry and the column maximum of the second entry of each payoff pair are starred we obtain the pattern

$(2^*,1)$	$(1,2^*)$
$(1,2^*)$	$(2^*,1)$

Since no payoff pair has both of its entries starred it follows that this game has no pure Nash equilibrium points.

The logic that underlies the Nash equilibrium is the expectation that as the players watch each other maneuver, the situation will naturally gravitate towards a Nash equilibrium outcome. Such was the case in the Cuban Missile Crisis. As those October days passed the Russians became convinced of the Americans' determination to persist and so they swerved. The diagram below demonstrates that this actual outcome is indeed one of the two Nash equilibrium points of that confrontation.

USSR

	Swerve	Persist
US Swerve	$(3,3)$	$(2^*,4^*)$
Persist	$(4^*,2^*)$	$(1,1)$

Prisoner's Dilemma has only one Nash equilibrium point with

	Refuse the deal	Accept the deal
Refuse the deal	$(-5,-5)$	$(-15,0^*)$
Accept the deal	$(0^*,-15)$	$(-10^*,-10^*)$

payoff pair $(-10,-10)$ which is consistent both with its maximin strategy and many observed solutions. Here, of course, the players cannot watch each other reach their decisions. Nevertheless, the oft reached outcome wherein both prisoners accept the deal does turn out to be a Nash equilibrium point.

The Nash equilibrium points have become a popular tool for theoretical economists. The next example should give the reader some feel for how this concept is used by them.

THE JOB APPLICANTS. *Firms 1 and 2 have one opening each for which they offer salaries 2a and 2b, respectively (the 2 is only used in order to prevent fractions from appearing later). Each of Ruth and Charlie can apply to only one of the positions and they must simultaneously decide whether to apply to firm 1 or to firm 2. If only one of them applies for a job, he gets it; if both apply for the same position, the firm hires one of them at random.*

This situation is modeled as the 2×2 nonzero-sum game

Charlie applies to

	firm 1	firm 2
Ruth applies to firm 1	(a,a)	$(2a,2b)$
firm 2	$(2b,2a)$	(b,b)

In this array the entries $(2a,2b)$ and $(2b,2a)$ are self-explanatory. The entry (a,a) is obtained by reasoning as follows. If both Ruth and Charlie apply to the same firm, then, because it was stipulated that the firm will select an applicant at random, each can expect a payoff of half the salary, i.e., a. A similar line of thought justifies the entry (b,b).

Not surprisingly, the analysis of the game depends on the relationship between a and b. Suppose first that

$$a \leq 2b \qquad \text{and} \qquad b \leq 2a.$$

In other words, suppose first that the two salaries are not too far out of line with each other in that neither exceeds double the other. Then the two entries $(2a, 2b)$ and $(2b, 2a)$ are both pure Nash equilibria. Such, for example might be the outcome if Ruth and Charlie became aware of each other's existence and came to some mutual agreement.

On the other hand, if the salaries are out of line with each other, say

$$b > 2a.$$

then it is the entry (b, b) that constitutes the unique pure Nash equilibrium point. This corresponds to both Ruth and Charlie applying for the better job. The case where $b = 2a$ is intermediary and has all of the above three outcomes as its Nash equilibria.

Thus, this highly simplified model predicts that if the disparity between the salaries is not too great then some reasonable distribution of the positions may happen. If one job is much better than the other, then both will apply for it resulting in a situation where one of them remains unemployed.

This model will be discussed again in greater depth in the next two chapters.

We mention in closing this chapter that the 2×2 nonzero-sum games have been classified in Rapoport, Guyer, and Gordon (1976) into 78 different types. This classification depends on the distribution of Pareto optimal outcomes, Nash equilibrium points, and their relationship to the maximin strategies.

Chapter Summary

Nonzero-sum games, where one player's gain is not necessarily the other player's loss are modeled as rectangular arrays of pairs of numbers. The pure maximin strategy can be used as a guideline in such games when they are played a single time. Two concrete games, Prisoner's Dilemma and Chicken were discussed in detail. Their analyses led to the definition of Pareto optimal and Nash equilibrium outcomes.

Chapter Terms

Better than	125	Chicken	125
Nash equilibrium	128	Nonzero-sum game	121
Pareto optimal outcome	125	Payoff pair	121
Principle of Rationality	122	Prisoner's dilemma	124
Pure maximin strategy	122	Pure value pair	122

EXERCISES 12

For each of the nonzero-sum games in Exercises 1–10 find the following:
   a) all the pure maximin strategies,

b) the pure maximin values,

c) all the pure value pairs,

d) all the Pareto optimal payoffs,

e) all the pure Nash equilibrium points,

f) the outcome of the game if Charlie is aware of Ruth's decision when he makes his, and Ruth knows of this,

g) the outcome of the game if Ruth is aware of Charlie's decision when she makes hers, and Charlie knows of this.

1.

$(2,3)$	$(1,4)$
$(0,5)$	$(4,1)$

2.

$(0,5)$	$(2,3)$
$(4,1)$	$(1,4)$

3.

$(1,3)$	$(3,1)$
$(2,2)$	$(2,5)$

4.

$(3,1)$	$(2,5)$
$(1,1)$	$(2,2)$

5.

$(5,4)$	$(4,1)$
$(1,4)$	$(3,2)$
$(1,1)$	$5,5)$

6.

$(1,2)$	$(3,4)$
$(5,6)$	$(7,8)$
$(7,6)$	$(5,5)$

7.

$(9,0)$	$(8,1)$	$(7,2)$
$(6,3)$	$(5,4)$	$(4,5)$
$(3,6)$	$(2,7)$	$(1,9)$

8.

$(9,0)$	$(8,-1)$	$(7,-2)$
$(6,-3)$	$(5,-4)$	$(4,-5)$
$(3,-6)$	$(2,-7)$	$(1,-9)$

9.

$(1,2)$	$(1,-2)$	$(1,1)$	$(0,3)$
$(2,1)$	$(0,0)$	$(-3,0)$	$(1,-2)$
$(3,-1)$	$(-2,1)$	$(2,1)$	$(2,2)$
$(2,1)$	$(2,2)$	$(1,-1)$	$(-1,1)$

10.

$(-1,2)$	$(-2,1)$	$(2,0)$	$(1,2)$
$(1,1)$	$(3,0)$	$(0,-3)$	$(-2,1)$
$(1,-2)$	$(3,0)$	$(1,2)$	$(0,3)$
$(3,0)$	$(2,2)$	$(-1,1)$	$(-1,1)$

Describe the following situations as nonzero sum games.

11. **Leader** (A. Coleman) Two motorists are waiting to enter a heavy stream of traffic from opposite ends of an intersection, and both are in a hurry to get to their destinations. When a gap in the traffic occurs, each must decide

whether to concede the right of way to the other or to drive into the gap. (Use the preferential ranking of the game of Chicken to assign numerical values to the payoffs).

12. **Battle of the sexes** (Luce and Raiffa) Ruth and Charlie, who are happily married, are planning an evening's entertainment. Ruth would like to go to the concert at the Arts Center, and Charlie would rather stay home and watch the ball game on TV. Still, both would rather spend their time together than separately. (Use the preferential ranking of the game of Chicken to assign numerical values to the payoffs).

13. **Two leagues** (Dixit & Nalebuff) Two football leagues, USFL and NFL, are deciding whether to schedule their games in the fall or in the spring. They estimate that 10 million viewers will watch football in the fall, but only 5 million will watch in the spring. If one league has a monopoly during a season, it gets the entire market. If both leagues schedule their games for the same season, the NFL gets 70% and the USFL gets 30% of the market.

14. **Oil Cartel** (Dixit & Nalebuff) Ruth and Charlie are the rulers of two countries that have formed an oil cartel. In order to keep the price of oil up, they have agreed to limit their productions respectively to 4 million and 1 million barrels per day. For each, cheating means producing an extra 1 million barrels each day. Depending on their decisions, their total output would therefore be 5, or 6, or 7 million barrels, with corresponding profit margins of $16, $12, and $8 per barrel.

15. **Robbery** (Dixit & Nalebuff) Ruth is typical homeowner, and Charlie an average burglar. Ruth is trying to decide whether to keep a gun in her home and Charlie faces the options of whether or not to bring a gun to his next break-in. (Use preferential ranking).

16. **Altruist's dilemma** (Heckathorn) Ruth and Charlie, who are married, are considering their Christmas gift strategy. To simplify matters, suppose they can each either spend a lot or a reasonable amount on each other's presents. Use preferential ranking to display this as a nonzero-sum game.

17. **Assurance game** (Heckathorn) Ruth and Charlie are the employees of a firm that has been remarkably successful over the last two years. They know that their boss has more than enough money to give them both a raise. Suppose each has the options of either not doing anything at all or else presenting their boss with the ultimatum: "If you don't give me a raise I quit!" Use preferential ranking to present this situation as a nonzero-sum game.

18. **Privileged game** (Heckathorn) This is essentially the same as Ex. 17 with the additional wrinkle that Ruth and Charlie are aware that when the boss caves in to one employee's ultimatum he will automatically give the other employee a smaller raise.

19. Does every nonzero-sum game have to have at least one Pareto optimal outcome? Justify your answer.

20. Give an example of a nonzero-sum game in which all the payoffs are distinct and each outcome is Pareto optimal.

13

MIXED STRATEGIES FOR
NONZERO-SUM GAMES

Mixed analogs of the Nash equilibrium and the maximin solutions
of 2 × 2 nonzero-sum games are offered. Examples from
economics and biology are discussed in detail.

Just like their zero-sum kin, nonzero-sum games may be repeated and their
strategies may be mixed. In analogy with the contents of the previous chapter
we shall discuss both mixed maximin strategies and mixed Nash equilibria.

As was done in the discussion of zero-sum games, we begin with 2×2 nonzero-sum games. A mixed strategy for the general 2×2 game

Charlie

(a_1, b_1)	(a_2, b_2)
(a_3, b_3)	(a_4, b_4)

Ruth

consists of a pair $[s, t]$ where s and t are nonnegative numbers such that

$$s + t = 1.$$

Ruth's and Charlie's general mixed strategies will again be denoted by $[1-p, p]$
and $[1 - q, q]$, respectively. It will prove convenient to refer to such a pair of
strategies as a *mixed strategy pair*. When Ruth and Charlie employ these mixed
strategies in a nonzero-sum game the expected payoff pair is computed in much
the same manner as it was for the zero-sum games, except that each player
focuses on his own payoffs. When the mixed strategies pair is $([1-p, p], [1-q, q])$,
Ruth's expected payoff is denoted by $e_R(p, q)$ and Charlie's expected payoff is
denoted by $e_C(p, q)$.

EXAMPLE 1. Compute the expected payoffs when Ruth and Charlie employ the mixed strategy pair $([.3, .7], [.6, .4])$ in the game

$(1, 2)$	$(3, -2)$
$(-1, 0)$	$(0, 4)$

The auxiliary diagram

	.6	.4
.3	$(1, 2)$	$(3, -2)$
.7	$(-1, 0)$	$(0, 4)$

yields

$$e_R(.7, .4) = .3 \times .6 \times 1 + .3 \times .4 \times 3 + .7 \times .6 \times (-1) + .7 \times .4 \times 0$$
$$= .18 + .36 - .42 + 0 = .12$$

and

$$e_C(.7, .4) = .3 \times .6 \times 2 + .3 \times .4 \times (-2) + .7 \times .6 \times 0 + .7 \times .4 \times 4$$
$$= .36 - .24 + 0 + 1.12 = 1.24.$$

In the previous chapter the pure maximin strategy of either player of the general 2×2 nonzero-sum game was obtained by each player confining his attention to his own portion of the payoff array, and such also is the case for the *mixed maximin strategy*. Each player will be assumed to be playing his array as a zero-sum game against an irrelevant opponent, and the mixed maximin strategy is that one which guarantees the best possible expected payoff. This payoff constitutes that player's *mixed maximin value* for that game.

EXAMPLE 2. Find the mixed maximin strategies and values for the game

$(5, 4)$	$(2, 2)$
$(4, 1)$	$(1, 3)$

In this game, Ruth's payoffs are

Ruth

5	2
4	1

which array, when viewed as a zero-sum game played by Ruth against an irrelevant opponent, has a saddle point in its upper right hand entry. Consequently,

Ruth's mixed maximin strategy is the pure strategy $[1, 0]$ and the mixed maximin value is 2. On the other hand, Charlie's payoffs constitute the array

Charlie

4	2
1	3

which is not strictly determined. Since his oddments are $[3 - 2, 4 - 1] = [1, 3]$, it follows that his mixed maximin strategy is $[.25, .75]$ and his mixed maximin value is

$$.25 \times 4 + .75 \times 2 = 2.5.$$

Since Ruth's and Charlie's portions of the payoff array of any nonzero-sum game are independent of each other, there will be in general no relationship between Ruth and Charlie's mixed maximin values. This, of course, stands in marked contrast to the state of affairs in zero-sum games where the player's mixed values are identical (or each other's negatives, depending on one's point of view). Thus, no analog of von Neumann's Theorem 6.6 exists for the mixed maximin strategies.

However, in 1950, John Nash succeeded in transferring the No Regrets aspect of von Neumann's Theorem to the nonzero-sum context. The significance of this achievement was underscored in 1994 when he was awarded the Nobel Prize for economics, in which discipline the concept of an equilibrium point has become an indispensable theoretical tool. The remainder of this chapter is devoted to a discussion of Nash's point of view together with several examples and applications. To formulate this new version it will again be assumed that the players are guided by the No Regrets policy enunciated in the previous chapter. In other words, after the repeated game is done, each player would like to rest assured that no other behavior on his part would have resulted in a better expected payoff. For the sake of completeness Nash's theorem is stated in its full generality even though its terms cannot be fully defined here. This general statement will be followed by a working restricted version.

NASH'S THEOREM 3. *Every n-person finite nonzero-sum game has an equilibrium point.*

The ensuing discussion and examples are confined mostly to two-person 2×2 nonzero-sum games. Our limited version of Nash's Theorem can now be stated as follows.

THEOREM 4. *In any 2×2 nonzero-sum game, there is a mixed strategy pair* $([1 - \bar{p}, \bar{p}], [1 - \bar{q}, \bar{q}])$ *such that*

$$e_R(\bar{p}, \bar{q}) \geq e_R(p, \bar{q}) \quad \text{for all } p, \quad 0 \leq p \leq 1 \tag{1}$$

and

$$e_C(\bar{p}, \bar{q}) \geq e_C(\bar{p}, q) \quad \text{for all } q, \quad 0 \leq q \leq 1. \tag{2}$$

Inequality (1) of this theorem says that if both players adopt the stipulated strategies then Ruth would have gained naught by adopting any other strategy. In other words, having employed the recommended strategy $[1 - \bar{p}, \bar{p}]$, Ruth has no cause for regrets as long as Charlie sticks to $[1 - \bar{q}, \bar{q}]$. Inequality (2) makes a similar assertion about her opponent: Charlie has no cause to regret his employment of $[1 - \bar{q}, \bar{q}]$ as long as Ruth sticks to $[1 - \bar{p}, \bar{p}]$. The pair of strategies $[1 - \bar{p}, \bar{p}], [1 - \bar{q}, \bar{q}]$ described by this theorem constitute a *mixed Nash equilibrium*. The numbers $e_R(\bar{p}, \bar{q})$ and $e_C(\bar{p}, \bar{q})$ are the game's *mixed value pair*. Together, the mixed Nash equilibrium and the mixed value pair are a *Nash equilibrium solution* of the game.

All known proofs of Nash's Theorem are nonconstructive. In other words, while they assert the existence of the mixed Nash equilibrium, they provide no method for actually deriving its constituent mixed strategies. Chapter 14 describes a method for finding the mixed Nash equilibria of 2×2 nonzero-sum games. The present chapter is confined to a description of a method for recognizing Nash equilibrium pairs. This topic is very similar to the issue of recognition of maximin and minimax strategies for symmetric zero-sum games discussed in Chapter 9.

First it is necessary to reconsider the optimal counterstrategies of Chapter 3 in the context of nonzero-sum games. Since each player is concerned only with his own payoffs, it follows that an *optimal counterstrategy* is one that maximizes a player's payoff. Consequently, Theorem 3.1 holds for nonzero-sum games as well (see Exercise 29):

THEOREM 5. *If one player of a nonzero-sum game employs a fixed strategy, then the opponent has an optimal counterstrategy that is pure.*

EXAMPLE 6. Suppose Ruth employs the strategy $[.2, .8]$ in the game

$(3, 2)$	$(2, 1)$
$(0, 3)$	$(4, 4)$

Find an optimal response for Charlie.

We know that Charlie has an optimal counterstrategy that is pure and so we compute the expected payoffs that correspond to the two pure strategies that are available to him. The auxiliary diagrams below contain Charlie's

	1	0
.2	2	1
.8	3	4

	0	1
.2	2	1
.8	3	4

payoffs only and they yield

$$.2 \times 2 + .8 \times 3 = 2.8 \qquad \text{for } [1, 0],$$
$$.2 \times 1 + .8 \times 4 = 3.4 \qquad \text{for } [0, 1].$$

Since Charlie is interested in maximizing his payoff, it follows that the pure strategy [0, 1] , which yields him a payoff of 3.4, is an optimal response for him.

EXAMPLE 7. Suppose Charlie employs the strategy [.7, .3] in the game

$(3, 2)$	$(2, 1)$
$(0, 3)$	$(4, 4)$

Find an optimal response for Ruth.

Since Ruth has an optimal counterstrategy that is pure, we compute the expected payoffs that correspond to the two pure strategies that are available to her. The auxiliary diagrams below contain Ruth's

	.7	.3
1	3	2
0	0	4

	.7	.3
0	3	2
1	0	4

payoffs only and they yield

$$.7 \times 3 + .3 \times 2 = 2.7 \qquad \text{for } [1, 0],$$
$$.7 \times 0 + .3 \times 4 = 1.2 \qquad \text{for } [0, 1].$$

Since Ruth is interested in maximizing her payoff, it follows that the pure strategy [1, 0] , which yields her a payoff of 2.7, is an optimal response for her.

We now turn to the issue of recognizing Nash equilibiria. Suppose your mathematical consultant has provided you, at cost, with a mixed Nash equilibrium pair for a game that is of interest to you. How can you be sure that the mathematician has not made a mistake and that you are indeed getting your money's worth? Theorem 5 above provides us with a straightforward method for checking on the proposed strategy pair. One need merely verify that none of the pure strategies at each player's disposal provide that player with a better payoff than that resulting from the proposed Nash equilibrium pair.

EXAMPLE 8. Verify that the mixed strategy pair ([.5, .5], [.4, .6]) constitutes a mixed Nash equilibrium for the game

$(3, 2)$	$(2, 1)$
$(0, 3)$	$(4, 4)$

Ruth's and Charlie's payoffs for the proposed Nash equilibrium pair are respectively,

	.4	.6
.5	$(3,2)$	$(2,1)$
.5	$(0,3)$	$(4,4)$

$$e_R(.5, .6) = .5 \times .4 \times 3 + .5 \times .6 \times 2 + .5 \times .4 \times 0 + .5 \times .6 \times 4$$
$$= .6 + .6 + 1.2 = 2.4,$$

and

$$e_C(.5, .6) = .5 \times .4 \times 2 + .5 \times .6 \times 1 + .5 \times .4 \times 3 + .5 \times .6 \times 4$$
$$= .4 + .3 + .6 + 1.2 = 2.5.$$

Is it possible for Ruth to improve her payoff by switching to an alternate strategy? If so, then there must be a pure alternate strategy that allows her to accomplish this. However, the auxiliary diagrams below both yield payoffs of

	.4	.6
1	$(3,2)$	$(2,1)$
0	$(0,3)$	$(4,4)$

	.4	.6
0	$(3,2)$	$(2,1)$
1	$(0,3)$	$(4,4)$

$$.4 \times 3 + .6 \times 2 = 2.4 \qquad \text{for } [1, 0],$$
$$.4 \times 0 + .6 \times 4 = 2.4 \qquad \text{for } [0, 1],$$

both of which equal the value of $e_R(.5, .6)$ computed above. This demonstrates that Ruth stands to gain nothing by switching to any alternate pure (or mixed) strategy. Thus, the proposed Nash equilibrium pair can cause Ruth no regrets.

Does Charlie have any cause for regrets? The auxiliary diagrams below yield payoffs

	1	0
.5	$(3,2)$	$(2,1)$
.5	$(0,3)$	$(4,4)$

	0	1
.5	$(3,2)$	$(2,1)$
.5	$(0,3)$	$(4,4)$

$$.5 \times 2 + .5 \times 3 = 2.5 \qquad \text{for } [1, 0],$$
$$.5 \times 1 + .5 \times 4 = 2.5 \qquad \text{for } [0, 1],$$

both of which equal the value of $e_C(.5, .6)$ computed above. Thus, Charlie has no cause for regrets either.

The conclusion is that the strategy pair $([.5, .5], [.4, .6])$ constitutes a Nash equilibrium pair for the given game.

The same method can also be used to recognize strategy pairs that are not Nash equilibria.

EXAMPLE 9. Decide whether $\left(\left[\frac{1}{3}, \frac{2}{3}\right], \left[\frac{1}{6}, \frac{5}{6}\right]\right)$ constitutes a Nash equilibrium pair for the game

$(5,1)$	$(0,0)$
$(0,0)$	$(1,5)$

The expected payoffs corresponding to the proposed strategy pair are

	$1/6$	$5/6$
$1/3$	$(5,1)$	$(0,0)$
$2/3$	$(0,0)$	$(1,5)$

$$e_R\left(\frac{2}{3}, \frac{5}{6}\right) = \frac{1}{3} \times \frac{1}{6} \times 5 + \frac{2}{3} \times \frac{5}{6} \times 1 = \frac{5 + 10}{18} = \frac{15}{18} = \frac{5}{6},$$

$$e_C\left(\frac{2}{3}, \frac{5}{6}\right) = \frac{1}{3} \times \frac{1}{6} \times 1 + \frac{2}{3} \times \frac{5}{6} \times 5 = \frac{1 + 50}{18} = \frac{51}{18} = \frac{17}{6}.$$

To see whether Ruth has any cause for regrets we compute the payoffs of the auxiliary diagrams below as

	$1/6$	$5/6$
1	$(5,1)$	$(0,0)$
0	$(0,0)$	$(1,5)$

	$1/6$	$5/6$
0	$(5,1)$	$(0,0)$
1	$(0,0)$	$(1,5)$

$$\frac{1}{6} \times 5 = \frac{5}{6} \qquad \text{for } [1, 0],$$

$$\frac{5}{6} \times 1 = \frac{5}{6} \qquad \text{for } [0, 1].$$

As these alternate payoffs equal the value of $e_R\left(\frac{2}{3}, \frac{5}{6}\right)$, Ruth stands to gain nothing by relinquishing the given strategy $\left[\frac{1}{3}, \frac{2}{3}\right]$. Passing on to Charlie, the auxiliary diagrams for the alternate pure strategies yield payoffs

	1	0
1/3	$(5,1)$	$(0,0)$
2/3	$(0,0)$	$(1,5)$

	0	1
1/3	$(5,1)$	$(0,0)$
2/3	$(0,0)$	$(1,5)$

$$\frac{1}{3} \times 1 = \frac{1}{3} \qquad \text{for } [1,0],$$
$$\frac{2}{3} \times 5 = \frac{10}{3} \qquad \text{for } [0,1].$$

Since the second of these, $\frac{10}{3} = 3.\overline{3}$, exceeds the value $\frac{17}{6} = 2.1\overline{6}$ of $e_C\left(\frac{2}{3}, \frac{5}{6}\right)$, it follows that Charlie could improve his payoff by abandoning the proposed strategy of $\left[\frac{1}{6}, \frac{5}{6}\right]$ in favor of the pure strategy $[0, 1]$. Hence, the given mixed strategy pair does not constitute a Nash equilibrium.

We offer the reader an alternate way of visualizing the defining properties of the Nash equilibrium. Suppose the values of $e_R(p,q)$ and $e_C(p,q)$, $p = 0, .2, .4, .6, .8, 1$, $q = 0, .2, .4, .6, .8, 1$ are tabulated for the game

$(3,1)$	$(1,5)$
$(1,2)$	$(4,1)$

in the form

p \ q	0	0.2	0.4	0.6	0.8	1
0	(3,1)	(2.6,1.8)	(2.2,2.6)	(1.8,3.4)	(1.4,4.2)	(1,5)
0.2	(2.6,1.2)	(2.4,1.8)	(2.2,2.4)	(2,3)	(1.8,3.6)	(1.6,4.2)
0.4	(2.2,1.4)	(2.2,1.8)	(2.2,2.2)	(2.2,2.6)	(2.2,3)	(2.2,3.4)
0.6	(1.8,1.6)	(2,1.8)	(2.2,2)	(2.4,2.2)	(2.6,2.4)	(2.8,2.6)
0.8	(1.4,1.8)	(1.8,1.8)	(2.2,1.8)	(2.6,1.8)	(3,1.8)	(3.4,1.8)
1	(1,2)	(1.6,1.8)	(2.2,1.6)	(2.8,1.4)	(3.4,1.2)	(4,1)

Observe that in the $q = .4$ column, the first entry of all of the payoff pairs has the constant value of 2.2, and that in the $p = .8$ row, the second entry has the constant value of 1.8. Consequently the mixed strategy pair $([.2, .8], [.6, .4])$ is a Nash equilibrium since neither player stands to gain anything by changing his mixed strategy.

While such a table can be a useful pedagogical tool, its construction is rather laborious and the details are greatly dependent on the exact numerical values of the constituent mixed strategies of the equilibrium. No more space will therefore be devoted to such tables in this text.

The next two examples are meant to demonstrate the wide range of applicability of the notion of a mixed Nash equilibrium. They come from the disciplines of economics and biology, respectively.

THE JOB APPLICANTS. In the previous chapter we modeled a situation wherein Ruth and Charlie were allowed to apply to one of two positions offering salaries $2a$ and $2b$ respectively, as the game

Charlie applies to

		firm 1	firm 2
Ruth applies to	firm 1	(a, a)	$(2a, 2b)$
	firm 2	$(2b, 2a)$	(b, b)

It was noted above that this game always has pure Nash equilibria whose exact values depend on the relative sizes of a and b. Specifically, if neither salary exceeds double the other, i.e., if

$$b \leq 2a \quad \text{and} \quad a \leq 2b,$$

then the outcomes corresponding to the payoffs $(2a, 2b)$ and $(2b, 2a)$ constitute pure Nash equilibria. In that case, however, the game also possesses a mixed Nash equilibrium (see Ex. 18) each of whose strategies is

$$\left[\frac{2a - b}{a + b}, \frac{2b - a}{a + b} \right].$$

The components of this common strategy can be interpreted as this model's predictions for the probabilities of Ruth (or Charlie) applying to the corresponding firm. I.e., if $a \leq 2b$ and $b \leq 2a$, then

$\dfrac{2a - b}{a + b}$ is the predicted probability of Ruth applying to firm 1

$\dfrac{2b - a}{a + b}$ is the predicted probability of Ruth applying to firm 2.

For example, if firms 1 and 2 offer salaries of \$20,000 and \$30,000 respectively, then $a = \$10,000$, $b = \$15,000$, and this model predicts that the probability of a player applying for firm 2's position is

$$\frac{2 \times 15,000 - 10,000}{10,000 + 15,000} = \frac{20,000}{25,000} = .8.$$

On the other hand, if either $a > 2b$ or $b > 2a$, or, in other words, if one of the salaries is more than double the other, then all the Nash equilibria are the pure ones that were already discussed in Chapter 12 .

This analysis provides economists with a starting point for an investigation of the question of what effect wage differentials have on the pool of applicants. The details of this investigation fall outside the scope of this book.

AN EVOLUTIONARY GAME. In recent years biological evolution has offered some applications of Nash equilibria as well. In many species mating is preceded by a duel between the males. Stags shove each other with their antlers and snakes have wrestling matches. The winner gets to mate. Similar intraspecies contests result from territorial disagreements. A surprising aspect of these contests is that nature seems to be pulling its punches. Some individuals run rather than fight. Stags do not gore each other's vulnerable sides and in some species snakes do not bite each other. This is, of course, quite sensible behavior for the species, and the biologists Maynard Smith and Price have constructed a game theoretic model for these contests in which this moderated ferocity is explained as a Nash equilibrium. An unusual feature of this game is that it pits the species in question against itself.

To define the game, we stipulate a species whose individuals engage in intraspecies duels. Each such confrontation constitutes a play of the game. The individual members of the species are classified as either *Hawks* or *Doves*. A Hawk always attacks in a confrontation and a Dove always runs away. The winner of a confrontation gets to mate, or gets the better territory, and so he is in a better position to propagate his genes. In the event of an actual physical struggle, the loser sustains an injury. The payoff to the species of a single play consists of the effect the duel has on the individual's ability to reproduce, an elusive quantity called his *fitness*. It will be assumed that the confrontation's winner's fitness is augmented by amount $2a$, and that a fight's loser's fitness is reduced by amount $2b$ (the 2 is introduced here into the payoffs just in order to simplify the subsequent calculations). The precise payoffs are computed as follows:

Hawk vs. Hawk: The two contestants continue fighting until one is injured. Inasmuch as each has a 50% chance of winning (and gaining $2a$) or losing (and losing $2b$), each of them is assigned the payoff

$$50\% \times 2a + 50\% \times (-2b) = a - b.$$

Hawk vs. Dove: Since the Dove runs away, no physical struggle takes place; Hawk gains $2a$ and Dove neither gains nor loses anything.

Dove vs. Dove: Again each contestant has a 50% chance of adding $2a$ to his fitness, but as there is no physical fight, there is no question of injury. Thus, each player is assigned the payoff

$$50\% \times 2a = a.$$

The resulting nonzero-sum game has the array

	Hawk	Dove
Hawk	$(a-b, a-b)$	$(2a, 0)$
Dove	$(0, 2a)$	(a, a)

$$(3)$$

In this game both Ruth and Charlie represent the same species, and each play consists of a confrontation between some individuals. The pure strategy $[1, 0]$

calls for the species to evolve Hawks only. The pure strategy $[0, 1]$ calls for the species to evolve Doves only. A mixed strategy consists of evolving a mixture of both Hawks and Doves within the species. Such a strategy would presumably be encoded into the species' genes. However, mutations do occur with high frequency, and it is reasonable to assume that whatever strategy is now in effect could not be improved upon by any mutations. After all, if a change (that is internal to the species) could improve on its overall fitness, then, given the amount of time most species have been in existence, such a change would have in all likelihood taken place a long time ago. Thus, it is reasonable to conclude that the current ratio of Hawks to Doves in this species is a stable quantity that maximizes its overall fitness. It will now be argued that the optimality of this ratio implies that it actually comes from a mixed Nash equilibrium of the nonzero-sum game (3).

Let us consider the mathematical analog of this optimality argument. Suppose the species is currently using the mixed strategy $[1-\bar{p}, \bar{p}]$, i.e., its Hawks-to-Doves ratio is now $(1-\bar{p}) : \bar{p}$. Each individual's confrontation then augments his fitness by the expected payoff computed from the auxiliary diagram

	$1 - \bar{p}$	$\bar{p}$
$1 - \bar{p}$	$(a - b, a - b)$	$(2a, 0)$
$\bar{p}$	$(0, 2a)$	(a, a)

This value is

$$e_R(\bar{p}, \bar{p}) = e_C(\bar{p}, \bar{p}) = (1 - \bar{p})^2(a - b) + \bar{p}(1 - \bar{p})2a + \bar{p}^2 a$$
$$= \bar{p}^2(a - b - 2a + a) + \bar{p}(-2(a - b) + 2a) + 1(a - b)$$
$$= -\bar{p}^2 b + 2\bar{p}b + a - b.$$

The reason $e_R(\bar{p}, \bar{p})$ and $e_C(\bar{p}, \bar{p})$ are equal is that both Ruth and Charlie represent the *same* species. The actual common value of these two expressions is in fact immaterial at this point. It is only necessary to keep in mind that $e_R(\bar{p}, \bar{p}) = e_C(\bar{p}, \bar{p})$.

Suppose now that a comparatively small subpopulation of this species mutates and begins breeding Hawks and Doves with a mixed strategy of $[1-p, p]$ for some $p \neq \bar{p}$. Since this subpopulation is presumed small, its individuals will mostly be confronting normal individuals, and hence the expected increment to the mutated individual's fitness from each battle is the common value of $e_R(p, \bar{p})$ and $e_C(\bar{p}, p)$ that correspond to the auxiliary diagrams

Normal population

		$1 - \bar{p}$	$\bar{p}$
	$1 - p$	$(a - b, a - b)$	$(2a, 0)$
Mutated population	p	$(0, 2a)$	(a, a)

or

Mutated population

		$1 - p$	p
Normal population	$1 - \bar{p}$	$(a - b, a - b)$	$(2a, 0)$
	$\bar{p}$	$(0, 2a)$	(a, a)

The normal individual, however, still confronts mostly other normal individuals and hence his fitness is augmented by the same quantity as before, namely,

$$e_R(\bar{p}, \bar{p}) = e_C(\bar{p}, \bar{p}).$$

The above Panglossian argument that mutated strategies will not improve upon the existing strategy entails, amongst others, the inequalities

$$e_R(\bar{p}, \bar{p}) \geq e_R(p, \bar{p}) \quad \text{and} \quad e_C(\bar{p}, \bar{p}) \geq e_C(\bar{p}, p).$$

These inequalities, however, are tantamount to the No Regrets guideline since, if we interpret game (3) as a game between the distinct opponents Ruth and Charlie, these inequalities say that neither Ruth nor Charlie could have improved their expected payoffs by changing from the strategy $[1 - \bar{p}, \bar{p}]$ to any other strategy $[1-p, p]$. Thus, the existing Hawks and Doves breeding strategy $[1-\bar{p}, \bar{p}]$ constitutes a mixed Nash equilibrium strategy of game (3). Let us examine the Nash equilibria of this game.

If $a > b$, then $a - b > 0$ and the outcome with payoff $(a - b, a - b)$ constitutes the only (pure or mixed) Nash equilibrium (Exercise 14.31). The associated pure strategy is $[1, 0]$. This can be interpreted as saying that if the advantage of winning outweighs the injuries that accompany a loss, then the species will breed only Hawks.

If $a < b$, then (Exercise 14.32) there are two pure Nash equilibria with respective payoffs $(2a, 0)$ and $(0, 2a)$ and another mixed Nash equilibrium with strategies

$$[1 - \bar{p}, \bar{p}] = [1 - \bar{q}, \bar{q}] = \left[\frac{a}{b}, 1 - \frac{a}{b}\right].$$

The pure Nash equilibria can be ignored since each has $[1, 0]$ as the strategy for one player and $[0, 1]$ as the other player's strategy - an impossible situation given that both players stand for one and the same species. This leaves us with the mixed Nash equilibrium as the one that describes the species' Hawks-to-Doves ratio. Note that as a diminishes in comparison to b, this model predicts that the species will breed fewer and fewer Hawks. In particular, if $a = 1$ and $b = 3$ i.e., if the effect of an injury from a duel outweighs the benefits of winning by a factor of 3-to-1, then this model predicts that only

$$\frac{a}{b} = \frac{1}{3}$$

of the individuals of the species will be Hawks.

Given the inherently inaccurate nature of the payoffs, the final case, $a = b$, because of the exact equality it demands, is of course unlikely to occur in reality. Exercise 14.34 asserts that in this case Hawks alone will be bred.

An alternate interpretation of the Nash-equilibrium strategy $[1 - \bar{p}, \bar{p}]$ in this evolutionary game is that it constitutes a ferocity index. On a scale of 0 (Hawk) to 1 (Dove), the quantity s denotes the willingness to fight that evolution has encoded into the genes of that species. This could explain why, when fighting each other, the aforementioned snakes wrestle rather than bite.

$m \times n$ nonzero-sum games. All the concepts developed here for 2×2 nonzero-sum games apply to larger games as well. An $m \times n$ nonzero-sum game is a rectangular array of m rows and n columns in which each entry is a pair of numbers. The entry in the ith row and jth column is denoted by $(a_{i,j}, b_{i,j})$. Given two mixed strategy pair $([p_1, p_2, \ldots, p_m], [q_1, q_2, \ldots, q_n])$, Ruth and Charlie's expected payoffs, denoted by e_R and e_C respectively are computed as

$$e_R = \text{sum of all } p_i \times q_j \times a_{i,j} \quad i = 1, 2, \ldots, m, j = 1, 2, \ldots, n,$$
$$e_C = \text{sum of all } p_i \times q_j \times b_{i,j} \quad i = 1, 2, \ldots, m, j = 1, 2, \ldots, n.$$

EXAMPLE 10. Compute the expected payoffs when Ruth and Charlie use the mixed strategy pair $([.5, .2, .3], [.1, .2, .3, .4])$ in the nonzero-sum game

$(1,2)$	$(0,-1)$	$(3,1)$	$(-2,0)$
$(1,-3)$	$(0,0)$	$(2,1)$	$(-1,1)$
$(3,2)$	$(1,1)$	$(-1,1)$	$(3,-1)$

The auxiliary diagram

	.1	.2	.3	.4
.5	$(1,2)$	$(0,-1)$	$(3,1)$	$(-2,0)$
.2	$(1,-3)$	$(0,0)$	$(2,1)$	$(-1,1)$
.3	$(3,2)$	$(1,1)$	$(-1,1)$	$(3,-1)$

yields the expected payoffs

$$\begin{aligned}
e_R &= .5 \times .1 \times 1 + .5 \times .2 \times 0 + .5 \times .3 \times 3 + .5 \times .4 \times (-2) \\
&\quad + .2 \times .1 \times 1 + .2 \times .2 \times 0 + .2 \times .3 \times 2 + .2 \times .4 \times (-1) \\
&\quad + .3 \times .1 \times 3 + .3 \times .2 \times 1 + .3 \times .3 \times (-1) + .3 \times .4 \times 3 \\
&= .05 + .45 - .4 + .02 + .12 - .08 + .09 + .06 - .09 + .36 = .58,
\end{aligned}$$

$$e_C = .5 \times .1 \times 2 + .5 \times .2 \times (-1) + .5 \times .3 \times 1 + .5 \times .4 \times 0$$
$$+ .2 \times .1 \times (-3) + .2 \times .2 \times 0 + .2 \times .3 \times 1 + .2 \times .4 \times 1$$
$$+ .3 \times .1 \times 2 + .3 \times .2 \times 1 + .3 \times .3 \times 1 + .3 \times .4 \times (-1)$$
$$= .1 - .1 + .15 - .06 + .06 + .08 + .06 + .06 + .09 - .12 = .32.$$

Theorem 5 offers the same guidance in the search for optimality in this wider context as well.

EXAMPLE 11. If Ruth employs the strategy [.5, .2, .3] in the game of Example 10, find an optimal counterstrategy for Charlie.

The auxiliary diagrams

	1	0	0	0
.5	$(1, 2)$	$(0, -1)$	$(3, 1)$	$(-2, 0)$
.2	$(1, -3)$	$(0, 0)$	$(2, 1)$	$(-1, 1)$
.3	$(3, 2)$	$(1, 1)$	$(-1, 1)$	$(3, -1)$

	0	1	0	0
.5	$(1, 2)$	$(0, -1)$	$(3, 1)$	$(-2, 0)$
.2	$(1, -3)$	$(0, 0)$	$(2, 1)$	$(-1, 1)$
.3	$(3, 2)$	$(1, 1)$	$(-1, 1)$	$(3, -1)$

	0	0	1	0
.5	$(1, 2)$	$(0, -1)$	$(3, 1)$	$(-2, 0)$
.2	$(1, -3)$	$(0, 0)$	$(2, 1)$	$(-1, 1)$
.3	$(3, 2)$	$(1, 1)$	$(-1, 1)$	$(3, -1)$

	0	0	0	1
.5	$(1, 2)$	$(0, -1)$	$(3, 1)$	$(-2, 0)$
.2	$(1, -3)$	$(0, 0)$	$(2, 1)$	$(-1, 1)$
.3	$(3, 2)$	$(1, 1)$	$(-1, 1)$	$(3, -1)$

yield the following payoffs for Charlie:

$$e_C = .5 \times 2 + .2 \times (-3) + .3 \times 2 = 1 - .6 + .6 = 1 \quad \text{for } [1,0,0,0],$$
$$e_C = .5 \times (-1) + .2 \times 0 + .3 \times 1 = -.5 + .3 = -.2 \quad \text{for } [0,1,0,0],$$
$$e_C = .5 \times 1 + .2 \times 1 + .3 \times 1 = .5 + .2 + .3 = 1 \quad \text{for } [0,0,1,0],$$
$$e_C = .5 \times 0 + .2 \times 1 + .3 \times (-1) = .2 - .3 = -.1 \quad \text{for } [0,0,0,1].$$

Since 1 is the largest of these payoffs, it follows that both $[1, 0, 0, 0]$ and $[0, 0, 1, 0]$ are optimal counterstrategies for Charlie.

The procedure used to check on proposed Nash equilibrium pairs for 2×2 games can be applied in the more general context of $m \times n$ games as well.

EXAMPLE 12. Show that $([.4, .4, .2], [.2, .6, 0, .2])$ is a Nash equilibrium for the nonzero-sum game

$(0,1)$	$(1,0)$	$(1,0)$	$(0,1)$
$(1,0)$	$(0,1)$	$(1,0)$	$(2,-1)$
$(1,0)$	$(1,0)$	$(1,0)$	$(-1,2)$

The auxiliary diagram

	.2	.6	0	.2
.4	$(0,1)$	$(1,0)$	$(1,0)$	$(0,1)$
.4	$(1,0)$	$(0,1)$	$(1,0)$	$(2,-1)$
.2	$(1,0)$	$(1,0)$	$(1,0)$	$(-1,2)$

yields the expected payoffs

$$e_R = .4 \times .6 \times 1 + .4 \times .2 \times 1 + .4 \times .2 \times 2 + .2 \times .2 \times 1 + .2 \times .6 \times 1$$
$$+ .2 \times .2 \times (-1) = .24 + .08 + .16 + .04 + .12 - .04 = .6,$$
$$e_C = .4 \times .2 \times 1 + .4 \times .2 \times 1 + .4 \times .6 \times 1 + .4 \times .2 \times (-1) + .2 \times .2 \times 2$$
$$= .08 + .08 + .24 - .08 + .08 = .4.$$

On the other hand, if Charlie sticks to his $[.2, .6, 0, .2]$ and Ruth experiments with her pure strategies, she gets the payoffs

$$.6 \times 1 = .6 \qquad \qquad \text{for } [1,0,0],$$
$$.2 \times 1 + .2 \times 2 = .6, \qquad \text{for } [0,1,0],$$
$$.2 \times 1 + .6 \times 1 + .2 \times (-1) = .6 \quad \text{for } [0,0,1],$$

all of which equal the value of e_R. Thus, Ruth has no reason to abandon her given strategy. If Ruth sticks to her strategy of $[.4, .4, .2]$ and Charlie experiments with his pure strategies, he gets the payoffs

$$
\begin{aligned}
.4 \times 1 &= .4 & \text{for } [1,0,0,0], \\
.4 \times 1 &= .4 & \text{for } [0,1,0,0], \\
0 & & \text{for } [0,0,1,0], \\
.4 \times 1 + .4 \times (-1) + .2 \times 2 &= .4 & \text{for } [0,0,0,1],
\end{aligned}
$$

none of which is better than the value of e_C (note that one is actually worse). Thus, Charlie has no reason to regret staying with the given strategy either, and so the given strategy pair is indeed a Nash equilibrium.

Chapter Summary

The notion of a mixed Nash equilibrium point of a nonzero-sum game was defined. Nash's existence theorem was stated for 2×2 games and a method for the verification of Nash equilibiria was provided. Applications to theoretical economics and biology were discussed.

Chapter Terms

Dove	144	e_C	147
$e_C(p,q)$	135	Equilibrium solution	138
e_R	147	$e_R(p,q)$	135
Evolutionary game	144	Hawk	144
Job applicants	143	Mixed Nash equilibrium	138
Mixed strategy pair	135		

EXERCISES 13

Compute the expected payoff when the games of Exercises 1–6 are played with the specified mixed strategy pairs.

1.
$(4,4)$	$(0,1)$
$(1,0)$	$(1,1)$

$([.7, .3], [.2, .8])$

2.
$(4,1)$	$(0,2)$
$(1,1)$	$(1,0)$

$([.2, .8], [.5, .5])$

3.
$(4,0)$	$(0,1)$
$(1,2)$	$(1,5)$

$([.3, .7], [.8, .2])$

4.

(4, 3)	(0, 2)
(1, 0)	(1, 1)

$([.7, .3], [0, 1])$

5.

(4, 2)	(0, 1)
(1, 2)	(1, 0)

$([1, 0], [0, 1])$

6.

(1, 4)	(1, 1)
(2, 0)	(0, 1)

$([.1, .9], [.5, .5])$

For each of the games in Exercises 7–12:
   a) Determine the mixed maximin strategies and values;
   b) Decide whether the specified mixed strategies pair constitutes a Nash equilibrium.

7.

(3, 2)	(2, 4)
(2, 3)	(4, −3)

$\left(\left[\frac{1}{2}, \frac{1}{2}\right], \left[\frac{2}{3}, \frac{1}{3}\right]\right)$

8.

(3, 2)	(2, 4)
(2, 3)	(4, −3)

$\left(\left[\frac{3}{4}, \frac{1}{4}\right], \left[\frac{1}{2}, \frac{1}{2}\right]\right)$

9.

(3, 2)	(2, 4)
(2, 3)	(4, −3)

$\left(\left[\frac{3}{4}, \frac{1}{4}\right], \left[\frac{2}{3}, \frac{1}{3}\right]\right)$

10.

(2, −3)	(−1, 3)
(0, 1)	(1, −2)

$\left(\left[\frac{1}{3}, \frac{2}{3}\right], \left[\frac{1}{3}, \frac{2}{3}\right]\right)$

11.

(2, −3)	(−1, 3)
(0, 1)	(1, −2)

$\left(\left[\frac{1}{3}, \frac{2}{3}\right], \left[\frac{1}{2}, \frac{1}{2}\right]\right)$

12.

(2, −3)	(−1, 3)
(0, 1)	(1, −2)

$\left(\left[\frac{3}{4}, \frac{1}{4}\right], \left[\frac{1}{3}, \frac{2}{3}\right]\right)$

13. Decide whether the strategy pair $\left(\left[\frac{1}{8}, 0, \frac{5}{8}, \frac{1}{4}\right], \left[\frac{1}{2}, \frac{1}{2}, 0\right]\right)$ constitutes a Nash equilibrium for the nonzero-sum game

$(1,1)$	$(0,2)$	$(-1,3)$
$(0,2)$	$(-1,3)$	$(2,0)$
$(1,1)$	$(0,2)$	$(1,1)$
$(-1,3)$	$(2,0)$	$(0,2)$

14. Decide whether the strategy pair $\left(\left[\frac{1}{4}, \frac{1}{4}, 0, \frac{1}{2}\right], \left[\frac{1}{2}, 0, \frac{1}{4}, \frac{1}{4}\right]\right)$ constitutes a Nash equilibrium for the nonzero-sum game

$(1,0)$	$(0,1)$	$(1,0)$	$(-1,2)$
$(1,0)$	$(1,0)$	$(-1,2)$	$(1,0)$
$(-1,2)$	$(1,0)$	$(1,0)$	$(1,0)$
$(0,1)$	$(1,0)$	$(1,0)$	$(1,0)$

15. Decide whether the strategy pair $\left(\left[0, \frac{2}{7}, \frac{2}{7}, \frac{3}{7}\right], \left[\frac{2}{7}, \frac{3}{7}, \frac{2}{7}, 0\right]\right)$ constitutes a Nash equilibrium for the nonzero-sum game

$(-1,2)$	$(1,0)$	$(0,1)$	$(1,0)$
$(1,0)$	$(-1,2)$	$(1,0)$	$(0,1)$
$(0,1)$	$(1,0)$	$(-1,2)$	$(1,0)$
$(1,0)$	$(-1,2)$	$(1,0)$	$(-1,2)$

16. Decide whether the strategy pair $\left(\left[\frac{2}{7}, 0, \frac{2}{7}, \frac{3}{7}\right], \left[0, \frac{2}{7}, \frac{3}{7}, \frac{2}{7}\right]\right)$ constitutes a Nash equilibrium for the nonzero-sum game

$(-1,2)$	$(1,0)$	$(0,1)$	$(1,0)$
$(1,0)$	$(-1,2)$	$(1,0)$	$(0,1)$
$(0,1)$	$(1,0)$	$(-1,2)$	$(1,0)$
$(1,0)$	$(-1,2)$	$(1,0)$	$(-1,2)$

17. Decide whether the strategy pair $\left(\left[\frac{2}{7}, 0, \frac{2}{7}, \frac{3}{7}\right], \left[\frac{2}{7}, \frac{3}{7}, \frac{2}{7}, 0\right]\right)$ constitutes a Nash equilibrium for the nonzero-sum game

$(-1,2)$	$(1,0)$	$(0,1)$	$(1,0)$
$(1,0)$	$(-1,2)$	$(1,0)$	$(0,1)$
$(0,1)$	$(1,0)$	$(-1,2)$	$(1,0)$
$(1,0)$	$(-1,2)$	$(1,0)$	$(-1,2)$

18*. Show that the strategy pair $\left(\left[\frac{2a-b}{a+b}, \frac{2b-a}{a+b}\right], \left[\frac{2a-b}{a+b}, \frac{2b-a}{a+b}\right]\right)$ constitutes a mixed Nash equilibrium for the Job Applicants game whenever neither salary exceeds double the other.

19. Suppose $a = \$12,000$ and $b = \$25,000$ in the Job Applicants game. What is the predicted probability of Ruth applying to the position offered by
   a) Firm 1?   b) Firm 2?

20. Suppose $a = \$24,000$ and $b = \$20,000$ in the Job Applicants game. What is the predicted probability of Ruth applying to the position offered by
   a) Firm 1?   b) Firm 2?

21. Suppose $a = \$10,000$ and $b = \$12,000$ in the Job Applicants game. What is the predicted probability of Ruth applying to the position offered by
   a) Firm 1?   b) Firm 2?

22. Suppose $a = \$50,000$ and $b = \$20,000$ in the Job Applicants game. What is the predicted probability of Ruth applying to the position offered by
   a) Firm 1?   b) Firm 2?

23. Suppose $a = \$12,000$ and $b = \$18,000$ in the Job Applicants game. What is the predicted probability of Ruth applying to the position offered by
   a) Firm 1?   b) Firm 2?

24. Suppose $a = \$11,000$ and $b = \$26,000$ in the Job Applicants game. What is the predicted probability of Ruth applying to the position offered by
   a) Firm 1?   b) Firm 2?

25. Suppose $a = 2$ and $b = 3$ in the Maynard-Smith & Price Evolutionary Game. What is the predicted proportion of
   a) Hawks?   b) Doves?

26. Suppose $a = 3$ and $b = 2$ in the Maynard-Smith & Price Evolutionary Game. What is the predicted proportion of
   a) Hawks?   b) Doves?

27. Suppose $a = 15$ and $b = 9$ in the Maynard-Smith & Price Evolutionary Game. What is the predicted proportion of
   a) Hawks?   b) Doves?

28. Suppose $a = 6$ and $b = 15$ in the Maynard-Smith & Price Evolutionary Game. What is the predicted proportion of
   a) Hawks?   b) Doves?

29*. Prove Theorem 5.

14

FINDING MIXED NASH EQUILIBRIA FOR 2 x 2 NONZERO-SUM GAMES

A graphical method for finding the mixed Nash equilibria of 2×2 nonzero-sum games is described.

All known proofs of Nash's Theorem are nonconstructive. In other words, while they assert the existence of the mixed Nash equilibrium, they provide no method for actually deriving its constituent mixed strategies. However, a graphical method for finding these equilibria for 2×2 games is known and will now be presented. For this purpose it is convenient to have an explicit formula for the expected payoff when a given mixed strategy pair is used. The subsequent discussion and eventual proof pertain to the game

Charlie

Ruth	(a_1, b_1)	(a_2, b_2)
	(a_3, b_3)	(a_4, b_4)

The following statement is easily verified by a reference to the appropriate auxiliary diagram (Exercise 28).

If Ruth and Charlie are employing mixed strategy pairs $([1-p, p], [1-q, q])$ *in the general* 2×2 *nonzero-sum game of, then*

$$e_R(p, q) = (a_1 - a_2 - a_3 + a_4)pq + (a_3 - a_1)p + (a_2 - a_1)q + a_1$$

and (1)

$$e_C(p, q) = (b_1 - b_2 - b_3 + b_4)pq + (b_3 - b_1)p + (b_2 - b_1)q + b_1.$$

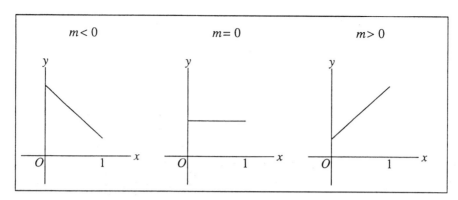

FIGURE 14.1. Looking for a Maximum.

The solution we are about to describe is based on the following observations regarding maximum values.

LEMMA 1. *For any fixed m, c, and any variable x, $0 \le x \le 1$, the value of $mx + c$ is maximized by*

only	$x = 0$	*if*	$m < 0,$	
any	$0 \le x \le 1$	*if*	$m = 0,$	
only	$x = 1$	*if*	$m > 0.$	

Figure 1 contains typical graphs of the straight line segment $y = mx + c$, $0 \le x \le 1$, in each of the three cases described in Lemma 1, and the justification of this assertion follows from the fact that the maximum value of $mx + c$ corresponds to the highest point on the graph of $y = mx + c$.

The Nash equilibrium points of 2×2 nonzero-sum games will be found as the intersections of two graphs. One of these graphs describes all the mixed strategy pairs which will bring no regrets to Ruth, and the other graph describes all the mixed strategy pairs which bring no regrets to Charlie.

RUTH'S NO REGRETS GRAPH. *Which mixed strategy pairs $([1 - p, p], [1 - q, q])$ result in no regrets for Ruth?* Her expectation when an arbitrary pair $([1 - p, p], [1 - q, q])$ is employed in the general game is

$$e_R(p, q) = (a_1 - a_2 - a_3 + a_4)pq + (a_3 - a_1)p + (a_2 - a_1)q + a_1.$$

Since the effect on the value of $e_R(p, q)$ of changing p depends on the coefficients of p, it makes sense to factor out the variable p wherever possible and write

$$e_R(p, q) = mp + c$$

where

$$m = (a_1 \doteq a_2 - a_3 + a_4)q + (a_3 - a_1), c = (a_2 - a_1)q + a_1.$$

It now follows from Lemma 1 that Ruth will have no regrets in the following three cases:

$$p = 0 \text{ if } m < 0 \quad | \quad 0 \le p \le 1 \text{ if } m = 0 \quad | \quad p = 1 \text{ if } m > 0. \tag{2}$$

The corresponding strategy pairs $([1 - p, p], [1 - q, q])$ can now be plotted in a Cartesian coordinate system with a horizontal p-axis and a vertical q-axis.

EXAMPLE 2. Draw Ruth's No Regrets graph for the game

$(3, 2)$	$(2, 4)$
$(2, 3)$	$(4, -3)$

Here

$$\begin{aligned} m &= (a_1 - a_2 - a_3 + a_4)q + (a_3 - a_1) \\ &= (3 - 2 - 2 + 4)q + (2 - 3) \\ &= 3q - 1. \end{aligned}$$

Here the three cases of (2) become

$p = 0$	$0 \le p \le 1$	$p = 1$
if $m < 0$	if $m = 0$	if $m > 0$
if $3q - 1 < 0$	if $3q - 1 = 0$	if $3q - 1 > 0$
if $3q < 1$	if $3q = 1$	if $3q > 1$
if $q < 1/3$	if $q = 1/3$	if $q > 1/3$

Keeping in mind that $0 \le q \le 1$, Ruth's No Regrets graph consists of the three line segments containing, respectively, all those points (p, q) such that

$p = 0$	$0 \le p \le 1$	$p = 1$
$0 \le q \le 1/3$	$q = 1/3$	$1/3 \le q \le 1$

A drawing of Ruth's No Regrets graph appears in Figure 2.

Ruth's No Regrets graph of Figure 2 should be interpreted as follows. The vertical segment along the y-axis tells us that Ruth should employ the pure strategy $[1, 0]$ whenever Charlie uses $q < 1/3$. The horizontal portion corresponds to the observation that if Charlie employs the mixed strategy $[2/3, 1/3]$ then Ruth's expected payoff does not depend on her strategy; she can expect the same payoff of

$$c = (a_2 - a_1)q + a_1 = (2 - 3) \times \frac{1}{3} + 3 = 2\frac{2}{3}$$

regardless of how she behaves. Finally, the vertical portion above $p = 1$ tells us that Ruth should employ the pure strategy $[0, 1]$ whenever Charlie uses $q > 1/3$.

CHARLIE'S NO REGRETS GRAPH. *Which mixed strategy pairs* $([1 - p, p], [1 - q, q])$ *result in no regrets for Charlie?* His expectation when an arbitrary pair $([1 - p, p], [1 - q, q])$ is employed in the general game is

$$e_C(p, q) = (b_1 - b_2 - b_3 + b_4)pq + (b_3 - b_1)p + (b_2 - b_1)q + b_1.$$

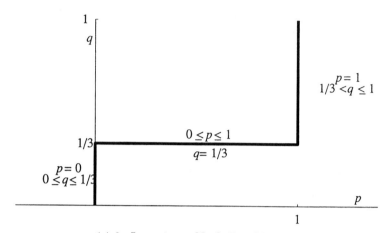

14.2. Locating a Nash Equilibrium.

Since the effect on the value of $e_C(p,q)$ of changing q depends on the coefficients of q, it makes sense to factor out the variable q wherever possible and write

$$e_C(p,q) = m'q + c'$$

where

$$m' = (b_1 - b_2 - b_3 + b_4)p + (b_2 - b_1),$$
$$c' = (b_3 - b_1)p + b_1.$$

It now follows from Lemma 1 that Charlie will have no regrets in the following three cases:

$$q = 0 \text{ if } m' < 0 \quad | \quad 0 \le q \le 1 \text{ if } m' = 0 \quad | \quad q = 1 \text{ if } m' > 0. \tag{3}$$

The mixed strategy pairs that cause Charlie no regrets can now be graphed in the same $p - q$ coordinate system that was used for Ruth's No Regrets graph.

EXAMPLE 3. Draw Charlie's No Regrets graph for the game

$(3, 2)$	$(2, 4)$
$(2, 3)$	$(4, -3)$

Here

$$m' = (b_1 - b_2 - b_3 + b_4)p + (b_2 - b_1)$$
$$= (2 - 4 - 3 + (-3))p + (4 - 2)$$
$$= -8p + 2.$$

The three cases of (3) now become:

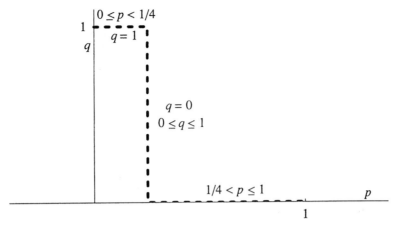

FIGURE 14.3. A No Regrets Graph for Charlie.

$q = 0$	$0 \leq q \leq 1$	$q = 1$
if $m' < 0$	if $m' = 0$	if $m' > 0$
if $-8p - 2 < 0$	if $-8p + 2 = 0$	if $-8p + 2 > 0$
if $-8p < -2$	if $-8p = -2$	if $-8p > -2$
if $p > 1/4$	if $p = 1/4$	if $p < 1/4$

Keeping in mind that $0 \leq p \leq 1$, Charlie's No Regrets graph consists of the three line segments containing, respectively, all those points (p, q) such that

$1/4 < p \leq 1$	$p = 1/4$	$0 \leq p < 1/4$
$q = 0$	$0 \leq q \leq 1$	$q = 1$

A drawing of Charlie's No Regrets graph appears in Figure 3.

Charlie's No Regrets graph of Figure 3 should be interpreted as follows. The short horizontal segment to the immediate right of the q axis tells us that Charlie should employ the pure strategy $[0, 1]$ whenever Ruth uses $p < 1/4$. The vertical portion corresponds to the observation that if Ruth employs the mixed strategy $[3/4, 1/4]$ then Charlie's expected payoff does not depend on his strategy; he can expect the same payoff of

$$c' = (b_3 - b_1)p + b_1 = (3 - 2) \times \frac{1}{4} + 2 = 2\frac{1}{4}.$$

regardless of how he behaves. Finally, the horizontal portion along the p-axis tells us that Charlie should employ the pure strategy $[1, 0]$ whenever Ruth uses $p > 1/4$.

Each player's No Regrets graph consists of those mixed strategy pairs $([1 - p, p], [1 - q, q])$ that put that player in the position of not having to regret his or her own decision. Since the Nash equilibrium points consist of those strategy pairs $([1 - p, p], [1 - q, q])$ that cause regrets to neither player, they constitute the intersections of Ruth's and Charlie's No Regrets graphs.

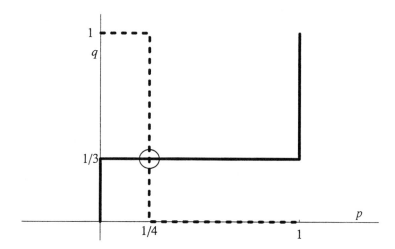

FIGURE 14.4. Locating a Nash Equilibrium.

EXAMPLE 4. To find the Nash equilibrium points of the game of Examples 2 and 3 it is now only necessary to superimpose the graphs of Figures 2 and 3 in Figure 4. Ruth's and Charlie's No Regrets graphs intersect in one (circled) point $(1/4, 1/3)$ which is this game's only Nash equilibrium point.

In this equilibrium Ruth will use the mixed strategy $[1 - \frac{1}{4}, \frac{1}{4}] = [\frac{3}{4}, \frac{1}{4}]$ and Charlie will use the mixed strategy $[1 - \frac{1}{3}, \frac{1}{3}] = [\frac{2}{3}, \frac{1}{3}]$. That these strategies do constitute a mixed Nash equilibrium can be verified directly:

$$e_R(p, \frac{1}{3}) = (3 - 2 - 2 + 4) \times p \times \frac{1}{3} + (2 - 3) \times p + (2 - 3) \times \frac{1}{3} + 3$$

$$= p - p - \frac{1}{3} + 3 = \frac{8}{3};$$

$$e_C(\frac{1}{4}, q) = (2 - 4 - 3 - 3) \times \frac{1}{4} \times q + (3 - 2) \times \frac{1}{4} + (4 - 2) \times q + 2$$

$$= -2q + \frac{1}{4} + 2q + 2 = \frac{9}{4}.$$

Thus, neither player stands to gain anything by relinquishing his Nash equilibrium strategy. It follows from these calculations that the expected payoff pair associated with this equilibrium is $(8/3, 9/4)$.

As the next example demonstrates, a game may possess both pure and mixed Nash equilibrium points.

EXAMPLE 5. Find the Nash equilibrium points of the game

$(3, 2)$	$(2, 1)$
$(0, 3)$	$(4, 4)$

To construct Ruth's No Regrets graph we compute

$$m = (3 - 2 - 0 + 4)q + (0 - 3) = 5q - 3.$$

It follows from (2) that on Ruth's No Regrets graph

$p = 0$	$0 \le p \le 1$	$p = 1$
if $m < 0$	if $m = 0$	if $m > 0$
if $5q - 3 < 0$	if $5q - 3 = 0$	if $5q - 3 > 0$
if $5q < 3$	if $5q = 3$	if $5q > 3$
if $q < 3/5$	if $q = 3/5$	if $q > 3/5$

Keeping in mind that $0 \le q \le 1$, Ruth's No Regrets graph consists of the three line segments containing, respectively, all those points (p, q) such that

$p = 0$	$0 \le p \le 1$	$p = 1$
$0 \le q < 3/5$	$q = 3/5$	$3/5 < q \le 1$

Ruth's No Regrets graph is the solid line of Figure 5.

To draw Charlie's No regrets graph we compute

$$m' = (2 - 1 - 3 + 4)p + (1 - 2) = 2p - 1.$$

It follows from (3) that on Charlie's No Regrets graph

$q = 0$	$0 \le q \le 1$	$q = 1$
if $m' < 0$	if $m' = 0$	if $m' > 0$
if $2p - 1 < 0$	if $2p - 1 = 0$	if $2p - 1 > 0$
if $2p < 1$	if $2p = 1$	if $2p > 1$
if $p < 1/2$	if $p = 1/2$	if $p > 1/2$

Keeping in mind that $0 \le p \le 1$, Charlie's No Regrets graph consists of the three line segments containing, respectively, all those points (p, q) such that

$0 \le p < 1/2$	$p = 1/2$	$1/2 < p \le 1$
$q = 0$	$0 \le q \le 1$	$q = 1$

This graph is depicted by the dashed line in Figure 5.

The Nash equilibrium points are given by the (circled) intersections of these two No Regrets graphs. Thus, there are three such equilibria corresponding to O: $(p = 0, q = 0)$, A: $(p = \frac{1}{2}, q = \frac{3}{5})$, and B: $(p = 1, q = 1)$, respectively. In other words, equilibrium O involves both players using the pure strategy [1, 0]; equilibrium A involves Ruth using the mixed strategy [.5, .5] and Charlie using the mixed strategy [.4, .6]; and in equilibrium B both players use the pure strategy [0, 1]. A glance at the given game will verify that O and B constitute pure Nash equilibria. That A is also an equilibrium point can be verified directly:

$$e_R(p, .6) = (3 - 2 - 0 + 4) \times p \times .6 + (0 - 3) \times p + (2 - 3) \times .6 + 3$$
$$= 3p - 3p - .6 + 3 = 2.4,$$
$$e_C(.5, q) = (2 - 1 - 3 + 4) \times .5 \times q + (3 - 2) \times .5 + (1 - 2) \times q + 2$$
$$= q + .5 - q + 2 = 2.5.$$

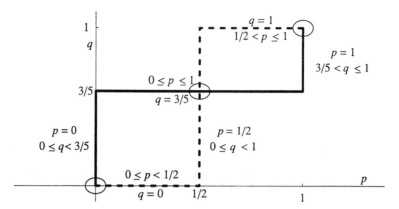

FIGURE 14.5. Locating a Nash Equilibrium.

Thus, neither player stands to gain anything by relinquishing the equilibrium strategy. It also follows from these calculations that the expected payoff pair associated with the mixed Nash equilibrium A is $(2.4, 2.5)$.

The graph of the next example looks quite different from those of the previous two.

EXAMPLE 6. Find the Nash equilibria of the game

$(1,4)$	$(2,2)$
$(2,2)$	$(4,1)$

To construct Ruth's No Regrets graph we compute

$$m = (1 - 2 - 2 + 4)q + (2 - 1) = q + 1.$$

It follows from (2) that on Ruth's No Regrets graph

$p = 0$	$0 \leq p \leq 1$	$p = 1$
if $m < 0$	if $m = 0$	if $m > 0$
if $q + 1 < 0$	if $q + 1 = 0$	if $q + 1 > 0$
if $q < -1$	if $q = -1$	if $q > -1$

Keeping in mind that $0 \leq q \leq 1$, the first two possibilities make no contribution to Ruth's No Regrets graph. Consequently this graph consists of the single line segment:

$$p = 1$$
$$0 \leq q \leq 1.$$

This is the solid vertical line segment in Figure 6.

To construct Charlie's No Regrets graph we compute

$$m' = (4 - 2 - 2 + 1)p + (2 - 4) = p - 2.$$

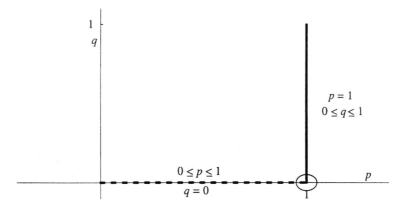

FIGURE 14.6. Locating a Nash Equilibrium.

It follows from (6) that on Charlie's No Regrets graph

$q = 0$	$0 \leq q \leq 1$	$q = 1$
if $m' < 0$	if $m' = 0$	if $m' > 0$
if $p - 2 < 0$	if $p - 2 = 0$	if $p - 2 > 0$
if $p < 2$	if $p = 2$	if $p > 2$

Keeping in mind that $0 \leq p \leq 1$, we see that the last two alternatives make no contributions to the graph. Consequently Charlie's No regrets graph consists of the line segment:

$$0 \leq p \leq 1$$
$$q = 0.$$

This graph is depicted as the dashed line in Figure 6.

The Nash equilibrium point is given by the (circled) intersection of the two graphs. At this point we have $p = 1$ and $q = 0$. Consequently, this Nash equilibrium point consists of the pure strategies $[0, 1]$ for Ruth and $[1, 0]$ for Charlie, with the payoff pair $(2, 2)$. It is interesting to note that the pure strategies $[1, 0]$ for Ruth and $[0, 1]$ for Charlie do not constitute an equilibrium even though they yield exactly the same payoff pair $(2, 2)$.

THE JOB APPLICANTS. In the previous two chapters we modeled a situation wherein Ruth and Charlie were allowed to apply to one of two positions offering salaries $2a$ and $2b$ respectively, as the game

		Charlie applies to	
		firm 1	firm 2
Ruth applies to	firm 1	(a, a)	$(2a, 2b)$
	firm 2	$(2b, 2a)$	(b, b)

It was noted above that this game always has pure Nash equilibria whose exact locations depend on the relative sizes of a and b. We now go on to find the mixed Nash equilibria. To construct Ruth's No Regrets graph we compute

$$m = (a - 2a - 2b + b)q + (2b - a) = -(a + b)q + (2b - a).$$

It follows from (2) that

$p = 0$	$0 \leq p \leq 1$
if $m < 0$	if $m = 0$
if $-(a + b)q + (2b - a) < 0$	if $-(a + b)q + (2b - a) = 0$
if $-(a + b)q < -(2b - a)$	if $-(a + b)q = -(2b - a)$
if $q > \frac{2b-a}{a+b}$	if $q = \frac{2b-a}{a+b}$

$p = 1$
if $m > 0$
if $-(a + b)q + (2b - a) > 0$
if $-(a + b)q > -(2b - a)$
if $q < \frac{2b-a}{a+b}$

If

$$a < 2b \text{ and } b < 2a \tag{4}$$

then (see Exercise 33)

$$0 < \frac{2b - a}{a + b} < 1.$$

In this case Ruth's No Regrets graph is depicted by the solid heavy line of Figure 7. Since Charlie's and Ruth's positions are interchangeable in this game, a similar argument (Exercise 36) yields the dashed heavy line of the same figure as Charlie's No Regrets graph.

Two of these Nash equilibria are pure and are identical with the ones already discussed in the previous chapter. The third one yields

$$\left[1 - \frac{2b - a}{a + b}, \frac{2b - a}{a + b} \right] = \left[\frac{2a - b}{a + b}, \frac{2b - a}{a + b} \right]$$

as the mixed Nash-equilibrium strategy for both Ruth and Charlie. The significance of the Nash equilibria of this game was already discussed above.

Proofs*

John Nash's important theorem asserts the existence of an equilibrium for all finite n-person (nonzero sum) games. Here we content ourself with the restricted cases we have been considering. The proof is based on the solution technique developed above.

THEOREM 7. *Every two person 2 × 2 nonzero sum game has a Nash Equilibrium point.*

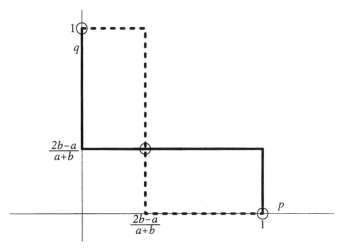

FIGURE 14.7. Nash Equilibria for Job Applicants.

PROOF. To prove the existence of a Nash Equilibrium point it suffices to show that Ruth's and Charlie's No Regrets graphs always intersect. We begin by listing all the possible forms that these graphs can take.

With reference to the 2×2 game G illustrated in the beginning of the chapter, set $D = a_1 - a_2 - a_3 + a_4$ and $E = a_3 - a_1$. Then the quantity that determines Ruth's No Regrets graph can be expressed as $m = Dq + E$. Consequently, Ruth's No Regrets graph is the intersection of the unit square with the union of the following three sets of points

$$\{(0, q) \text{ such that } Dq + E < 0]\}$$
$$\{(p, q) \text{ such that } 0 \leq p \leq 1 \text{ and } Dq + E = 0\}$$
$$\{(1, q) \text{ such that } Dq + E > 0\}.$$

Depending on the signs of D and E this union consists of one of the graphs in Figure 8, where it should be born in mind that the quantity $-E/D$ could fall *anywhere* on the q-axis. Consequently, depending on the value of $-E/D$, Ruth's No regrets graph has one of the forms listed in Figure 9. A similar argument whose details are left to the reader, or else an argument based on the symmetrical roles of p and q, permit us to conclude that Charlie's No Regrets graph has one of the forms listed in Figure 10.

In all cases Ruth's graph connects the horizontal sides of the unit square whereas Charlie's graph connects the vertical sides of the square. The existence of the required intersection therefore follows whenever either Ruth's graph contains an entire vertical side or else Charlie's graph contains an entire horizontal side. In the remaining four cases, where neither player's graph contains an entire side, the superimposition of the two graphs illustrated in Figure 11 makes it again clear that the required intersection exists. □

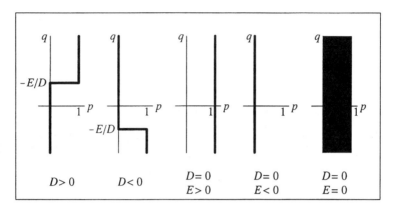

FIGURE 14.8. Ruth's Unrestricted No Regrets Line.

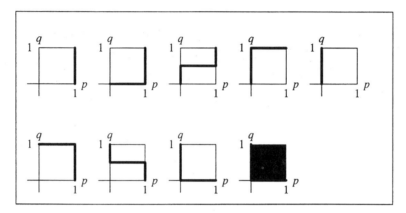

FIGURE 14.9. Ruth's No Regrets Graphs.

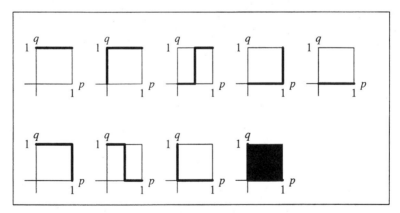

FIGURE 14.10. Charlie's No Regrets Graphs.

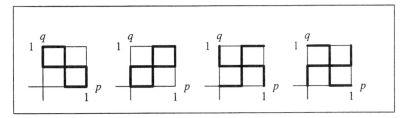

FIGURE 14.11. Locating the Nash Equilibria.

Chapter Summary

A graphical method for finding the Nash equilibria of 2×2 nonzero-sum games was described. This was followed by a proof of the existence of mixed Nash equilibria for such games.

Chapter Terms

Charlie's No Regrets Graph   157   Ruth's No Regrets graph   156

EXERCISES 14

For each of the games in Exercises 1–25:
- a) Determine the mixed maximin strategies and values;
- b) Find the mixed Nash equilibrium points and mixed value pairs.
- c) Whenever a nonpure Nash equilibrium $([1 - \bar{p}, \bar{p}], [1 - \bar{q}, \bar{q}])$ exists, verify that $e_R(\bar{p}, q)$ and $e_C(p, \bar{q})$ are independent of q and p respectively.

1.

$(4,4)$	$(0,1)$
$(1,0)$	$(2,3)$

2.

$(4,1)$	$(0,2)$
$(1,2)$	$(1,0)$

3.

$(4,0)$	$(0,1)$
$(1,2)$	$(2,5)$

4.

$(4,3)$	$(0,2)$
$(1,0)$	$(2,1)$

5.

$(4,2)$	$(0,1)$
$(1,2)$	$(1,0)$

6.

$(1,4)$	$(1,1)$
$(2,0)$	$(0,1)$

7.

$(3,4)$	$(2,2)$
$(1,3)$	$(3,4)$

8.

$(3,1)$	$(2,2)$
$(1,3)$	$(3,0)$

9.

$(1,3)$	$(1,2)$
$(2,0)$	$(4,1)$

10.

$(4,1)$	$(3,4)$
$(3,5)$	$(1,0)$

11.

$(0,4)$	$(2,1)$
$(1,0)$	$(5,1)$

12.

$(1,4)$	$(4,2)$
$(3,3)$	$(1,4)$

13.

$(3,1)$	$(6,2)$
$(3,3)$	$(1,1)$

14.

$(0,3)$	$(2,4)$
$(1,0)$	$(5,3)$

15.

$(1,1)$	$(4,4)$
$(3,3)$	$(1,0)$

16.

$(3,4)$	$(0,1)$
$(2,0)$	$(1,1)$

17.

$(3,1)$	$(0,2)$
$(2,1)$	$(1,0)$

18.

$(3,0)$	$(0,1)$
$(2,2)$	$(1,5)$

19.

$(3,3)$	$(2,2)$
$(2,0)$	$(1,1)$

20.

$(3,2)$	$(0,1)$
$(4,2)$	$(3,0)$

21.

$(2,4)$	$(2,1)$
$(1,0)$	$(0,1)$

22.

$(1,4)$	$(4,2)$
$(5,3)$	$(1,4)$

23.

$(1,4)$	$(4,2)$
$(2,3)$	$(1,1)$

24.

$(2,3)$	$(2,3)$
$(1,1)$	$(0,1)$

25.

$(1,1)$	$(4,4)$
$(5,3)$	$(1,0)$

26. Does the Leader game of Exercise 12.11 have a mixed Nash equilibrium point? If so, interpret it.
27. Does the Battle of the Sexes game of Exercise 12.12 have a mixed Nash equilibrium point? If so, interpret it.
28. Verify the equations of (1).
29. Verify directly that the points circled in Figure 7 are Nash equilibria of the Job Applicants game.
30. Verify that if the assumptions of (4) do not hold then all the Nash equilibrium points of the Job Applicants game are pure.
31. Verify that if $a > b$ then the evolutionary game has only one Nash equilibrium point and this point is pure.
32. Derive all the Nash equilibria of the evolutionary game under the assumption that $a < b$.
33. Show that if a and b are any two positive numbers such that $a < 2b$ and $b < 2a$, then
$$0 < \frac{2b - a}{a + b} < 1 \text{ and } 0 < \frac{2a - b}{a + b} < 1.$$
34. Show that if $a = b$ in the evolutionary game, then the model predicts that the species will evolve Hawks only.
35. Can a 2 × 2 game have exactly one pure and one nonpure Nash equilibria? Justify your answer.
36. Derive Charlie's No Regrets graph in Figure 7.

BIBLIOGRAPHY

Borel, E., Several papers dating 1921–1927 were translated into English by L. J. Savage and appear with commentaries by M. Fréchet and J. von Neumann in *Econometrica*, **21**(1953), 95–127.

Coleman, A., *Game Theory and Experimental Games*, Pergamon Press, Oxford, 1982.

Davenport, W. C., Jamaican Fishing Village, *Papers in Carribean Anthropology*, **59**, (I. Rouse ed.), Yale University Press, 1960.

Davis, M. D., *Game Theory*, Basic Books, New York, 1983.

Dixit, A. and Nalebuff B., *Thinking Strategically*, W. W. Norton & Company, New York 1991.

Dresher, M., *The Mathematics of Games and Strategy, Theory and Applications*, Dover, New York, 1981.

Gale, D., *The Theory of Linear Economic Models*, McGraw-Hill, New York, 1960.

Haywood, O. G. Jr., Military Decision and Game Theory, *J. of the Operations Reseach Society of America*, **2**(1954), 365–385.

Heckathorn, D. D., The Dynamics and Dilemmas of Collective Action, *American Sociological Review*, to appear.

Jones, A. J., *Game Theory: Mathematical Models of Conflict, Halstead Press*, Chichester England, 1980.

Luce, R. D. and Raiffa, H., *Games and Decision: Introduction and Critical Survey*, John Wiley and Sons, London, 1957.

Maynard Smith, J. and Parker, G. A., The Logic of Animal Conflict, *Nature*, **246**(1973), 15–18.

Maynard Smith, J., *Evolution and the Theory of Games*, Cambridge University Press, London, 1982.

Montgomery, J. D., Equilibrium Wage Dispersion and Interindustry Wage Differentials, *Quarterly Journal of Economics* 1991, 163–179.

Morris, P., *Introduction to Game Theory*, Springer-Verlag, New York 1994.

Nash, J., Non-cooperative games, *Annals of Mathematics*, **54**(1950), 286–295.

Packel, E., *The Mathematics of Games and Gambling*, The Mathematical Association of America, Washington, D.C., 1981.

Poundstone, W., *Prisoner's Dilemma*, Doubleday, New York, 1992.

Rapoport, A., Guyer, M. J., and Gordon, D. G., *The 2×2 Game*, The University of Michigan Press, Ann Arbor, 1976.

Sinervo, B. and Lively, C. M., The Rock-Paper-Scissors game and the evolution of alternative male strategies, *Nature*, **380**(1996), 240–243.

Straffin, P. D., *Game Theory and Strategy*, The Mathematical Association of America, Washington, D.C., 1993.

Thomas, L. C., Games, *Theory and Applications*, Ellis Horwood Limited, Chichester England, 1984.

Venttsel, Y. S., *Elements of Game Theory*, Mir Publishers, Moscow, 1980.

von Neumann, J., Zur Theorie der Gesellschaftsspiele, *Mathematische Annalen*, **100**(1928), 295–300.

von Neumann, J., and Morgenstern, O., *Theory of Games and Economic Behavior*, Princeton University Press, Princeton, 1947.

Williams, J. D., *The Complete Strategyst*, Dover, New York, 1986.

Zagare, F. C., *Game Theory: Concepts and Applications*, Sage Publications, Beverly Hills, 1984.

SOLUTIONS TO SELECTED EXERCISES

Chapter 1:
1. a) $144°$, $108°$, $72°$, $36°$ b) 0 c) .5 d) $-.8$ e) .1 f) (2, 3)
3. a) $36°$, $72°$, $108°$, $144°$ b) .5 c) $-.1$ d) -1.3 e) $-.6$ f) (1, 3)
5. a) $108°$, $252°$ b) 83% c) 85% d) 84% e) 83.8%

Chapter 2:
1. 2.94 **3.** 1.8 **5.** .37 **7.** -3.5 **9.** .12 **11.** -4

13.

		B-day	No
Frank	Flowers	1.5	1
	No	-10	0

15.

		Rain	Shine
Merill	Umbrellas	250	50
	Glasses	0	300

Chapter 3:
1. $[0, 1]$ **3.** $[0, 1]$ **5.** $[0, 1]$ **7.** $[0, 0, 0, 1]$
9. $[0, 0, 1, 0]$ **11.** $[0, 0, 1]$ **13.** $[0, 0, 1]$ **15.** $[1, 0, 0]$
17. $[0, 0, 0, 1, 0]$

Chapter 4:
1. b) $[1/3, 2/3]$ c) 5/3 d) $p \le 2/3$, $p \ge 2/3$
3. b) Any strategy c) 1 d) $[0, 1]$ for all p
5. b) $[0, 1]$ c) 4 d) $p \ge 1/3$, $p \le 1/3$
7. b) $[1, 0]$ c) 1 d) $[1, 0]$ for all p
9. b) $[1, 0]$ c) 2 d) $p \le 2/3$, $p \ge 2/3$
11. b) $[1, 0]$ c) 1 d) $[1, 0]$ for all p

13. b) $[7/13, 6/13]$ c) $11/13$ d) $p \geq 6/13$, $p \leq 6/13$
15. b) $[1, 0]$ c) 2   d) $[0, 1]$ for all p

Chapter 5:

  1. b) $[2/3, 1/3]$ c) $5/3$ d) $q \geq 1/3$, $q \leq 1/3$
  3. b) $[0, 1]$ c) 1   d) $[0, 1]$ for all q
  5. b) $[1, 0]$ c) 4   d) $[0, 1]$ for all q
  7. b) $[1, 0]$ c) 1   d) $[1, 0]$ for all q
  9. b) $[1, 0]$ c) 2   d) $[1, 0]$ for all q
11. b) Any strategy   c) 1   d) $[1, 0]$ for all q
13. b) $[4/13, 9/13]$ c) $11/13$ d) $q \leq 9/13$, $q \geq 9/13$
15. b) $[0, 1]$ c) 2   d) $[1, 0]$ for all q

Chapter 6:

  1. $[1/3, 2/3]$, $[2/3, 1/3]$, $5/3$ **3.** Any strategy, $[0, 1]$, 1
  5. $[0, 1]$, $[1, 0]$, 4   **7.** $[1, 0]$, $[1, 0]$, 1
  9. $[1, 0]$, $[1, 0]$, 2   **11.** $[1, 0]$, any strategy, 1
13. $[7/13, 6/13]$, $[4/13, 9/13]$, $11/13$ **15.** $[1, 0]$, $[0, 1]$, 2
17. $[1, 0]$, $[1, 0]$, -3 **19.** $[1, 0]$, $[1, 0]$, 0
21. $[1, 0]$, $[0, 1]$, -2 **23.** $[1, 0]$, $[1, 0]$, 2
25. $[3/7, 4/7]$, $[4/7, 3/7]$, $9/7$ **27.** $[0, 1]$, $[0, 1]$, 0
29. $[2/3, 1/3]$, $[2/3, 1/3]$, $-1/3$ **31.** $[1, 0]$, $[0, 1, 0]$, 1
33. $[0, 0, 1, 0]$, $[1, 0]$, 2   **35.** $[0, 1, 0]$, $[0, 1, 0, 0]$, -2
37. $[0, 0, 1, 0]$, $[0, 0, 1, 0]$, 0   **39.** $[1, 0, 0, 0, 0]$, $[0, 1, 0, 0]$, 1
41. $x \geq 1$ **43.** $x \geq 1$ **45.** $1 \leq x \leq 2$ **47.** All x
49. $x \leq 4$ **51.** No x

		Stranger	
		Heads	Tails
Steve	Heads	-20	30
	Tails	10	-20

53.

   Steve: $[3/8, 5/8]$, Stranger: $[5/8, 3/8]$, value: $-1/8$
55. Merrill should invest \$150 in umbrellas and \$100 in glasses.
59. Yes   **61.** Yes

Chapter 7:

  1. $[4/7, 3/7]$, $[4/7, 0, 3/7]$, $2/7$ **3.** $[2/3, 1/3]$, $[0, 2/9, 7/9]$, $1/3$
  5. $[1 - p, p]$ $1/4 \leq p \leq 4/9$, $[1, 0, 0]$, 0   **7.** $[2/3, 1/3]$, $[0, 1/9, 0, 8/9]$, $-13/3$
  9. $[1, 0]$, $[1, 0, 0, 0]$ or $[0, 0, 1, 0]$, 2   **11.** $[0, 1, 0]$, $[1 - q, q]$ for $1/4 \leq q \leq 2/3$, 1
13. $[0, 1, 0]$, $[1, 0]$, 2   **15.** $[0, 0, 1, 0]$, $[1, 0]$, 3
17. $[0, 4/5, 0, 1/5]$, $[9/10, 1/10]$, $16/5$ **19.** $[0, 1/3, 2/3, 0]$, $[2/5, 3/5]$, 1
21. $[.5, .5, 0, 0]$ or many others, $[.5, .5]$, 3
23. Any strategy, any strategy, 1

Chapter 8:
1. $[0, 6/7, 1/7]$, $[0, 2/7, 0, 5/7, 0]$, $16/7$ **3.** $[1, 0, 0]$, $[0, 0, 1, 0, 0]$, 2
5. $[0, 0, 0, 2/3, 1/3]$, $[0, 5/6, 1/6]$, $-1/3$ **7.** $[.5, .5, 0]$, $[0, .5, .5, 0, 0]$, 1.5
9. $[.5, .5, 0, 0, 0]$, $[.5, .5, 0, 0]$, 1.5

Chapter 9: 1. R1 **3.** R2 **5.** R1
7. $[1, 0, 0]$, $[1, 0, 0]$, 0 **9.** $[3/7, 2/7, 2/7]$, $[3/7, 2/7, 2/7]$, 0
13. i) $[0, y, 1 - y, 0]$ for $4/7 \geq y \geq 5/9$
 ii) $[x, 0, 0, 1 - x]$ for $5/8 \geq x \geq 3/7$
 iii) $[2w, 2/3 - 2w, 1/3 - w, w]$ for $1/3 \geq w \geq 0$
15. No.

Chapter 10:
 3. Bet/Bet, Call, 0
 5. If $a \geq b$: Bet/Bet, Call, $(a - b)/2$. If $a < b$: Bet/Fold, Call, 0
 7. Raise/See, Call/Fold, 0 **9.** Raise/Raise, Call/Fold, 3
11. If $a \leq b$: See/See, Call/Fold, 0. If $b \leq a$: Raise/Raise, Call/Fold, $3(a - b)/8$.
13. Raise/See or Raise/Raise, Call/Fold, 1
15. See/See, Call/Fold, 0 **17.** B/B, C/B, 0
19. If $a \leq b$: B/PF or PC/PF, C/B, $(b - a)/2$;
 If $a \geq b$: B/B or B/PC or PC/PC, C/B, 0.

Chapter 11:
 1. $[0, 1]$, 0, any strategy, 2 **3.** $[0, 1, 0]$, 1, $[1, 0]$, 1
 5. Any strategy, -2, $[0, 1, 0, 0]$, 1 **7.** $[0, 0, 0, 1]$, 1, $[0, 0, 0, 1, 0]$, 2
 9. $0 \leq .8 \leq 2$ **11.** $3 \leq 3 \leq 3$ **13.** $-1 \leq 13/11 \leq 3$
17. a) No b) $[0, 1]$ c) No d) $[3/8, 5/8]$ e) No f) No g) $[.5, .5]$
19. a) No b) No c) $[1, 0]$ d) $[5/12, 7/12]$ e) No f) $[0, 1]$ g) $[.5, .5]$

Chapter 12:
 1. a) $[1, 0]$, $[1, 0]$ b) 1, 3 c) $(2, 3)$ d) All outcomes e) None f) $[1, 0]$, $[0, 1]$, $(1, 4)$ g) $[1, 0]$, $[1, 0]$, $(2, 3)$
 3. a) $[0, 1]$, $[1, 0]]$ b) 2, 2 c) $(2, 2)$ d) $(3, 1)$, $(2, 5)$ e) None f) $[0, 1]$, $[0, 1]$, $(2, 5)$ g) $[0, 1]$, $[1, 0]$, $(2, 2)$
 5. a) $[1, 0, 0]$, $[1, 0]$ or $[0, 1]$ b) 4, 1 c) $(5, 4)$ or $(4, 1)$ d) $(5, 5)$ e) $(5, 4)$, $(5, 5)$ f) $[1, 0, 0]$, $[1, 0]$, $(5, 4)$ or $[0, 0, 1]$, $[0, 1]$, $(5, 5)$
 7. a) $[1, 0, 0]$, $[0, 0, 1]$ b) 7, 2 c) $(7, 2)$ d) All outcomes e) $(7, 2)$ f) $[1, 0, 0]$, $[0, 0, 1]$, $(7, 2)$ g) $[1, 0, 0]$, $[0, 0, 1]$, $(7, 2)$
 9. a) $[1, 0, 0, 0]$, $[1, 0, 0, 0]$ or $[0, 0, 1, 0]$ b) 0, -1 c) $(1, 2)$ or $(1, 1)$ d) $(0, 3)$, $(3, -1)$, $(2, 2)$ e) $(2, 2)$, $(2, 2)$ f) $[0, 1, 0, 0]$, $[1, 0, 0, 0]$, $(2, 1)$ or $[0, 0, 1, 0]$, $[0, 0, 0, 1]$, $(2, 2)$ or $[0, 0, 0, 1]$, $[0, 1, 0, 0]$, or $(2, 2)$ g) $[0, 0, 0, 1]$, $[0, 1, 0, 0]$, $(2, 2)$ or $[0, 0, 1, 0]$, $[0, 0, 0, 1]$, $(2, 2)$
19. Yes.

Chapter 13:
 1. $(.86, 1.36)$ **3.** $(1.66, 1.88)$ **5.** $(0, 1)$ **7.** No **9.** Yes **11.** Yes **13.** Yes

15. No **17.** Yes **19.** a) 0   b) 1   **21.** a) 4/11   b) 7/11   **23.** a) .2   b) .8
25. a) 2/3   b)1/3   **27.** a) 100% b) None

Chapter 14:

1. a) $[1/5, 4/5]$, $8/5$, $[1/3, 2/3]$, 2   b) $(p, q) = (0, 0)$, $(.5, .6)$, $(1, 1)$
3. a) $[.2, .8]$, 1.6, $[0, 1]$, 1   b) $(p, q) = (1, 1)$
5. a) $[0, 1]$, 1, $[1, 0]$, 2   b) $(p, q) = (0, 0)$
7. a) $[2/3, 1/3]$, $7/3$, $[2/3, 1/3]$, $10/3$ b) $(p, q) = (0, 0)$, $(2/3, 2/3)$, $(1, 1)$
9. a) $[0, 1]$, 2, $[0, 1]$, 1   b) $(p, q) = (1, 1)$
11. a) $[0, 1]$, 1, $[0, 1]$, 1   b) $(p, q) = (1, 1)$
13. a) $[1, 0]$, 3, $[1/3, 2/3]$, $5/3$ b) $(p, q) = (0, 1)$, $(p, 0)$ for $p \geq 1/3$
15. a) $[2/5, 3/5]$, $11/5$, $[2/3, 1/3]$, 2   b) $(p, q) = (0, 1)$, $(.5, .4)$, $(1, 0)$
17. a) $[0, 1]$, 1, $[1, 0]$, 1   b) $(p, q) = (.5, .5)$
19. a) $[1, 0]$, 2, $[0, 1]$, 1   b) $(p, q) = (0, 0)$
21. a) $[1, 0]$, 2, $[0, 1]$, 1   b) $(p, q) = (0, 0)$
23. a) $[.25, .75]$, 1.75, $[1, 0]$, 3   b) $(p, q) = (1, 0)$
25. a) $[4/7, 3/7]$, $19/7$, $[2/3, 1/3]$, 2   b) $(p, q) = (1, 0)$, $(1/2, 4/7)$, $(0, 1)$

INDEX